MAD IN AMERICA

MAD IN AMERICA

BAD SCIENCE, BAD MEDICINE, AND THE ENDURING MISTREATMENT OF THE MENTALLY ILL

Robert Whitaker

PERSEUS PUBLISHING

Library of Congress Card information is available.

ISBN: 0-7382-0385-8

Perseus Publishing books are available at special discounts for bulk purchases in the U.S. by corporations, institutions, and other organizations. For more information, please contact the Special Markets Department at the Perseus Books Group, 11 Cambridge Center, Cambridge, MA 02142, or call (800) 255-1514 or (617)252-5298.

Perseus Publishing is a member of the Perseus Books Group
Text design by Trish Wilkinson
Set in 11-point New Baskerville by Perseus Books Group

1 2 3 4 5 6 7 8 9 10   03 02 01
First printing, December 2001

Find Perseus Publishing on the World Wide Web at
http://www.perseuspublishing.com

To my parents

Grateful acknowledgment is made for permission to reprint from: Walter Freeman and James Watts, *Psychosurgery,* second edition, 1950. Courtesy of Charles C. Thomas, Publisher, Ltd., Springfield Illinois.

"We are still mad about the mad. We still don't understand them and that lack of understanding makes us mean and arrogant, and makes us mislead ourselves, and so we hurt them."

—David Cohen

CONTENTS

PREFACE

MY INTEREST IN this subject, the history of medical treatments for the mad, began in a simple manner. In the summer of 1998, I stumbled onto an unusual line of psychiatric research, which I reported on for the *Boston Globe.* In order to study the "biology" of schizophrenia, American scientists were giving the mentally ill chemical agents—amphetamines, ketamine, and methylphenidate—expected to heighten their psychosis. That seemed an unusual thing to do, particularly since some of the people recruited into the experiments had come stumbling into emergency rooms seeking help. Equally striking was the response of "ex-patients" to the experiments.

They were outraged, but not particularly *surprised.*

That seemed more than a little curious—why would they not be surprised?—and then I bumped into several studies in the medical literature that really struck me as odd. Over the past twenty-five years, outcomes for people in the United States with schizophrenia have *worsened.* They are now no better than they were in the first decades of the twentieth century, when the therapy of the day was to wrap the insane in wet sheets. Even more perplexing, schizophrenia outcomes in the United States and other developed countries today are much worse than in the poor countries of the world. The World Health Organization has looked at this question repeatedly—initially, nobody could believe this disparity in outcomes—

and each time it has come back with the same result. Suffer a psychotic break in a poor country like India or Nigeria, and chances are that in a couple of years you will be doing fairly well. But suffer a similar break in the United States or other developed countries, and it is likely that you will become chronically ill. Why should that be so? Why should living in a country with rich resources, and with advanced medical treatments for disorders of every kind, be so toxic to those who are severely mentally ill? Or to put it another way, why should living in countries where the poor struggle every day to find enough to eat and treatment for a mental disorder is likely to be provided by a shaman, whose armamentarium may consist of witch-doctor potions, be so helpful to recovery?

This medical failure is a profound one. More than 2 million Americans suffer from schizophrenia, and their difficult lives bring unimaginable heartache to their families, and to others who love them. Too many of the people so diagnosed end up in prison, homeless, or shuttling in and out of psychiatric hospitals. Our society as a whole is affected by this failure as well, and in a way that we don't normally appreciate. We usually think of the financial burden: Schizophrenia, it is said, is a "disease" that costs the United States more than $45 billion annually. But there is a much deeper cost. We, as a society, are *estranged* from the "mad" in our midst. We fear them and their illness. We read of occasional acts of violence committed by those said to be schizophrenic, and we respond by setting up programs that focus on keeping them medicated. But is that the best response? If the medications work so well, then why do "schizophrenics" fare so poorly in the United States?

The search to understand this therapeutic failure necessarily takes one deep into history. The past becomes a foil for understanding the present. It is a journey that begins with the founding of the first hospital in the colonies by Pennsylvania Quakers in 1751, and from there one can trace a path, however winding and twisted, to the poor outcomes of today. It is also a history that contains one surprise after another. For instance, we think of the 1800s as a time when the insane were routinely chained up and neglected, and yet, in the early nineteenth century, there arose a form of humanitarian care that has never been equaled since. Go forward one hundred years, however, and the path detours into one of the darkest chapters in America's history, one that, I believe, we have never dared to fully explore. Yet it is in that dark chapter that one finds the seeds for today's failure.

What one also quickly discovers is that a history of mad medicine reveals very little about what it is like to be "crazy" or "insane," or, as we say today, "ill with schizophrenia." However, it does reveal a great deal about the society that would "cure" these patients. Medical treatments for the severely mentally ill inevitably reflect the societal and philosophical values of the day. What is the nature of man? What does it mean to be human? Where is the line between "normals" and the "mad" to be drawn? What rights do the "mentally ill" have over their own minds? The medical treatments a society employs all arise from its answers to those questions. As such, mad medicine does provide a prism through which to view a society, and that is why the poor outcomes for those diagnosed with schizophrenia raise questions, I would think, for all of us.

Robert Whitaker

one of the darkest chapters in American [illegible] that Lock[illegible]

[illegible] extreme [illegible] grand deal

[illegible] that [illegible] is the [illegible] followers.

What [illegible] directly [illegible]

[illegible] little about what [illegible]

[illegible], if [illegible]. However, it does [illegible]

[illegible] about the [illegible] that would carry [illegible]

[illegible] ments for the [illegible]

[illegible] social and philosophical values of the age. What is the [illegible]

[illegible] What does it mean to be human? [illegible]

[illegible]

Robert [illegible]

ACKNOWLEDGMENTS

THE SEED FOR this book was planted in the spring of 1998, when I met Vera Sharav at her home in New York City. She headed up Circare, a group composed primarily of parents of mentally ill children, and in the manner of a good journalist, she had used Freedom of Information requests to dig up documents that told of abuses in psychiatric research. Those documents sparked my curiosity, and so did her passion. She had the conviction of someone set on righting a wrong, and such conviction was infectious. I would never have done this book if I had not met her. Today, Vera heads a group called the Alliance for Human Research Protection.

That fall, I reported on psychiatric research for the *Boston Globe.* I owe a great deal of thanks to Dolores Kong, my collaborator on that series, and to Nils Bruzelius, its editor.

It was David Oaks, a psychiatric "survivor" and editor of the journal *Mind Freedom,* who then challenged me to look into the merits of modern drug treatments for schizophrenia. He did so in the manner of throwing down the gauntlet: Would I really be willing to investigate this? Three years later, I can say that I'm deeply grateful for his having done so. Wesley Alcorn, who in 1998 was president of NAMI's (National Alliance for the Mentally Ill) consumer council, similarly urged me to take a longer look at care of the mentally ill in this country.

Like anyone who writes a history, I went to school by reading the works of others who've written on the topic. In particular, I owe an intellectual debt to the following scholars: Andrew Scull, Nancy Tomes, Gerald Grob, Daniel Kevles, Allan Chase, Barry Mehler, Edward Shorter, Elliot Valenstein, Joel Braslow, Jack Pressman, Leonard Roy Frank, Mary Boyle, David Cohen, Peter Breggin, and Ann Braden Johnson. I am also grateful to Loren Mosher, who provided me with a wealth of documents related to the Soteria Project; similarly, Leonard Roy Frank provided me with a great deal of material on electroshock. A number of patients (or their parents) let me review personal and legal documents pertaining to their psychiatric care; Shalmah Prince, in particular, provided me with a detailed written record of her experience in psychiatric research.

Kevin Lang, my agent at Bedford Book Works, was instrumental in helping me shape my book proposal. My editor at Perseus, Amanda Cook, is every writer's dream: She let me loose to tell the story I wanted to tell, and then after I'd turned in a first draft, she took out her editor's pencil and in numerous ways big and small showed me how to improve the narrative and polish the text.

Finally, none of this would have been possible without the loving support of my wife, Andrea, who is forever patient with me, and whose reviews of the earliest drafts of the book were invaluable. And daily I counted my blessings for having three wonderful children, Rabi, Zoey, and Dylan.

PART ONE

THE ORIGINAL BEDLAM

(1750–1900)

1

BEDLAM IN MEDICINE

Terror acts powerfully upon the body, through the medium of the mind, and should be employed in the cure of madness.

—Benjamin Rush[1]

A VISITOR TO THE "mad" wards of Pennsylvania Hospital at the turn of the nineteenth century would have found the halls astir with an air of reform. A few years earlier, in 1796 to be exact, the lunatics had been moved from unheated, dingy cells in the basement, where they had often slept on straw and been confined in chains, to a new wing, where their rooms were above ground. Here the winter chill was broken by a coal-fired stove, and occasionally the mad patients could even take a warm bath. Most important of all, they now began to receive regular medical treatments—a regimen of care, physician Benjamin Rush proudly told the Pennsylvania Hospital overseers, that had "lately been discovered to be effectual in treating their disorder."[2]

The introduction of medical treatments had been a long time coming. In 1751, when Quakers and other community leaders in Philadelphia had petitioned the Pennsylvania colonial assembly for funds to build the hospital, the first in the colonies, they had told of medical care that could help restore sanity to the mad mind. "It has been found," wrote Benjamin Franklin, who authored the plea, "by the experience of many Years, that above two Thirds of the Mad People received into Bethlehem Hospital [in England] and there treated properly, have been perfectly cured."[3] English mad-doctors had indeed begun making such claims and had even published books describing their effective treatments. However, while Franklin and his fellow Quakers may have hoped to bring such medicine to the colonies, they also had a second reason for building the hospital. There were, they wrote, too many lunatics "going at large [who] are a Terror to their neighbors, who are daily apprehensive of the Violences they may commit." Society needed to be protected from the insane, and it was this second function—hospital as jail—that had taken precedence when the hospital opened in 1756.

In those early years, the lunatics were kept in gloomy, foul-smelling cells and were ruled over by "keepers" who used their whips freely. Unruly patients, when not being beaten, were regularly "chained to rings of iron, let into the floor or wall of the cell . . . restrained in hand-cuffs or ankle-irons," and bundled into Madd-shirts that "left the patient an impotent bundle of wrath."[4] A visiting reverend, Manasseh Cutler, described the sorry scene:

> We next took a view of the Maniacs. Their cells are in the lower story, which is partly underground. These cells are about ten feet square, made as strong as a prison . . . Here were both men and women, between twenty and thirty in number. Some of them have beds; most of them clean straw. Some of them were extremely fierce and raving, nearly or quite naked; some singing and dancing; some in despair; some were dumb and would not open their mouths.[5]

The lunatics also had to suffer the indignity of serving as a public spectacle. After the hospital opened, visiting the mad had quickly become a popular Sunday outing, similar to visiting a zoo. Philadelphians were eager to get a glimpse of these wretched creatures, with good sport on occasion to be had by taunting them, particularly those restrained in irons and easily roused into a rage. So frequent were the public's visits, and so disturbing to the insane, that the hospital managers erected a fence in 1760 "to prevent the Disturbance which is given to the Lunatics confin'd in the Cells by the great Numbers of People who frequently resort and converse with them."[6] But even an iron fence couldn't keep the public at bay, and so in 1762, the hospital, trying to make the best of an unfortunate situation, began charging a visitor's fee of four pence.

All of this began to change once Rush arrived at the hospital in 1783.

The lunatics could not have hoped for a more kind-hearted man to be their advocate. Born of Quaker parents, Rush was constantly championing liberal, humanitarian reforms. As a young man, he had been a member of the Continental Congress and a signer of the Declaration of Independence. He'd advocated for the abolition of slavery and prison reform, and he brought this same compassion to his treatment of the mad. At his request, the hospital's governing board built a new wing for the insane patients, which was completed in 1796, and soon many patients were enjoying the comforts of rooms furnished with hair mattresses and feather beds. Those who were well behaved were allowed to stroll about the hospital grounds and engage in activities like sewing, gardening, and cutting straw. Rush also believed that games, music, and friendship could prove helpful, and the hospital even agreed to his request that "a Well qualified Person be employed as a Friend and Companion to the Lunatics."[7] The insane, he explained to hospital attendants, needed to be treated with kindness and respect. "Every thing necessary for their comfort should be provided for them, and every promise made to them should be faithfully and punctually performed."[8]

But such humanitarian care could only go so far. Rush was also a man of science. He'd studied at the University of Edinburgh, the most prestigious medical school in the world at the time. There, he'd been mentored by the great William Cullen, whose *First Lines of the Practice of Physic* was perhaps the leading medical text of the day. The European mad-doctors had developed a diverse array of therapeutics for curing madness, and Rush, eager to make Pennsylvania Hospital a place of modern medicine, employed their methods with great vigor. And this was treatment of an altogether different type.

They Are Brutes, Aren't They?

One of the first English physicians to write extensively on madness, its nature, and the proper treatments for it was Thomas Willis. He was highly admired for his investigations into the nervous system, and his 1684 text on insanity set the tone for the many medical guides that would be written over the next 100 years by English mad-doctors. The book's title neatly summed up his view of the mad: *The Practice of Physick: Two Discourses Concerning the Soul of Brutes.* His belief—that the insane were animal-like in kind—reflected prevailing conceptions about the nature of man. The great English scientists and philosophers of the seventeenth century—Francis Bacon, Isaac Newton, John Locke, and others—had all argued that reason was the faculty that elevated humankind above the animals. This was the form of intelligence that enabled man to scientifically know his world, and to create a civilized society. Thus the insane, by virtue of having lost their reason, were seen as having descended to a brutish state. They were, Willis explained, fierce creatures who enjoyed superhuman strength. "They can break cords and chains, break down doors or walls . . . they are almost never tired . . . they bear cold, heat, watching, fasting, strokes, and wounds, without any sensible hurt."[9] The mad, he added, if they were to be cured, needed to hold their physicians in awe and think of them as their "tormentors."

> Discipline, threats, fetters, and blows are needed as much as medical treatment . . . Truly nothing is more necessary and more effective for the recovery of these people than forcing them to respect and fear intimidation. By this method, the mind, held back by restraint, is induced to give up its arrogance and wild ideas and it soon becomes meek and orderly. This is why maniacs often recover much sooner if they are treated with tortures and torments in a hovel instead of with medicaments.[10]

A medical paradigm for treating the mad had been born, and eighteenth-century English medical texts regularly repeated this basic wisdom. In 1751, Richard Mead explained that the madman was a brute who could be expected to "attack his fellow creatures with fury like a wild beast" and thus needed "to be tied down and even beat, to prevent his doing mischief to himself or others."[11] Thomas Bakewell told of how a maniac "bellowed like a wild beast, and shook his chain almost constantly for several days and nights . . . I therefore got up, took a hand whip, and gave him a few smart stripes upon the shoulders . . . He disturbed me no more."[12] Physician Charles Bell, in his book *Essays on the Anatomy of Expression in Painting*, advised artists wishing to depict madmen "to learn the character of the human countenance when devoid of expression, and reduced to the state of lower animals."[13]

Like all wild animals, lunatics needed to be dominated and broken. The primary treatments advocated by English physicians were those that physically weakened the mad—bleeding to the point of fainting and the regular use of powerful purges, emetics, and nausea-inducing agents. All of this could quickly reduce even the strongest maniac to a pitiful, whimpering state. William Cullen, reviewing bleeding practices, noted that some advised cutting into the jugular vein.[14] Purges and emetics, which would make the mad patient violently sick, were to be repeatedly administered over an extended period. John Monro, superintendent of Bethlehem Asylum, gave one of his patients sixty-one vomit-inducing emetics in six months, including strong doses on eighteen successive nights.[15] Mercury and other chemical agents, meanwhile, were used to induce nausea so fierce that that the patient could not hope to have the mental strength to rant and rave. "While nausea lasts," George Man Burrows advised, "hallucinations of long adherence will be suspended, and sometimes be perfectly removed, or perhaps exchanged for others, and the most furious will become tranquil and obedient." It was, he added, "far safer to reduce the patient by nauseating him than by depleting him."[16]

A near-starvation diet was another recommendation for robbing the madman of his strength. The various depleting remedies—bleedings, purgings, emetics, and nausea-inducing agents—were also said to be therapeutic because they inflicted considerable pain, and thus the madman's mind became focused on this sensation rather than on his usual raving thoughts. Blistering was another treatment useful for stirring great bodily pain. Mustard powders could be rubbed on a shaved scalp, and once the blisters formed, a caustic rubbed into the blisters to further irritate and infect the scalp. "The suffering that attends the formation of these pustules is often indescribable," wrote one physician. The madman's pain could be expected to increase as he rubbed his hands in the caustic and touched his genitals, a pain that would enable the patient to "regain consciousness of his true self, to wake from his supersensual slumber and to stay awake."[17]

All of these physically depleting, painful therapies also had a psychological value: They were feared by the lunatics, and thus the mere threat of their employment could get the lunatics to behave in a better manner. Together with liberal use of restraints and an occasional beating, the mad would learn to cower before their doctors and attendants. "In most cases it has appeared to be necessary to employ a very constant impression of fear; and therefore to inspire them with the awe and dread of some particular persons, especially of those who are to be constantly near them," Cullen wrote. "This awe and dread is therefore, by one means or other, to be acquired; in the first place by their being the authors of all the restraints that may be occasionally proper; but sometimes it may be necessary to acquire it even by stripes and blows. The former, although having the appearance of more severity, are much safer than strokes or blows about the head."[18]

Such were the writings of the English mad-doctors in the 1700s. The mad were to be tamed. But were such treatments really curative? In the beginning, the mad-doctors were hesitant to boldly make that claim. But gradually they began to change their tune, and they did so for a simple reason: It gave them a leg up in the profitable madhouse business.

Merchants of Madness

In eighteenth-century England, the London asylum Bethlehem was almost entirely a place for the poor insane. The well-to-do in London shipped their family lunatics to private madhouses, a trade that had begun to emerge in the first part of the century. These boarding homes also served as convenient dumping grounds for relatives who were simply annoying or unwanted. Men could get free from their wives in this manner—had not their noisome, bothersome spouses gone quite daft in the head? A physician who would attest to this fact could earn a nice sum—a fee for the consultation and a referral fee from the madhouse owner. Doctors who owned madhouses made out particularly well. William Battie, who operated madhouses in Islington and Clerkenwell, left an estate valued at between £100,000 and £200,000, a fabulous sum for the time, which was derived largely from this trade.[19]

Even though most of the mad and not-so-mad committed to the private madhouses came from better families, they could still expect neglect and the harsh flicker of the whip. As reformer Daniel Defoe protested in 1728, "Is it not enough to make any one mad to be suddenly clap'd up, stripp'd, whipp'd, ill fed, and worse us'd?"[20] In the face of such public criticism, the madhouse operators protested that their methods, while seemingly harsh, were remedies that could restore the mad to their senses. They weren't just methods for managing lunatics, but curative medical treatments. In 1758, Battie wrote: "Madness is, contrary to the opinion of some unthinking persons, as manageable as many other distempers, which are equally dreadful and obstinate."[21] He devoted a full three chapters to cures.

In 1774, the English mad trade got a boost with the passage of the Act for Regulating Madhouses, Licensings, and Inspection. The new law prevented the commitment of a person to a madhouse unless a physician had certified the person as insane (which is the origin of the term "certifiably insane"). Physicians were now the sole arbiters of insanity, a legal authority that made the mad-doctoring trade more profitable than ever. Then, in 1788, King George III suffered a bout of madness, and his recovery provided the mad-doctors with public proof of their curative ways.

Francis Willis, the prominent London physician called upon by the queen to treat King George, was bold in proclaiming his powers. He boasted to the English Parliament that he could reliably cure "nine out of ten" mad patients and that he "rarely missed curing any [patients] that I had so early under my care: I mean radically cured."[22] On December 5, 1788, he arrived at the king's residence in Kew with an assistant, three keepers, a straight waistcoat, and the belief that a madman needed to be broken like a "horse in a manège." King George III was so appalled by the sight of the keepers and the straight waistcoat that he flew into a rage—a reaction that caused Willis to immediately put him into the confining garment.

As was his custom, Willis quickly strove to assert his dominance over his patient. When the king resisted or protested in any way, Willis had him "clapped into the straight-waistcoat, often with a band across his chest, and his legs tied to the bed." Blisters were raised on the king's legs and quickly became infected, the king pleading that the pustules "burnt and tortured him"—a complaint that earned him yet another turn in the straight waistcoat. Soon his legs were so painful and sore that he couldn't walk, his mind now wondering how a "king lay in this damned confined condition." He was repeatedly bled, with leeches placed on his temples, and sedated with opium pills. Willis also surreptitiously laced his food with emetics, which made the king so violently sick that, on one occasion, he "knelt on his chair and prayed that God would be pleased either to restore Him to his Senses, or permit that He might die directly."

In the first month of 1789, the battle between the patient and doctor became ever more fierce. King George III—bled, purged, blistered, restrained, and sedated, his food secretly sprinkled with a tartar emetic to make him sick—sought to escape, offering a bribe to his keepers. He would give them annuities for life if they would just free him from the mad-doctor. Willis responded by bringing in a new piece of medical equipment—a restraint chair that bound him more tightly than the straight waistcoat—and by replacing his pages with strangers. The king would no longer be allowed the sight of familiar faces, which he took as evidence "that Willis's men meant to murder him."

In late February, the king made an apparently miraculous recovery. His agitation and delusions abated, and he soon resumed his royal duties. Historians today believe that King George III, rather than being mad, suffered from a rare genetic disorder called porphyria, which can lead to high levels of toxic substances in the body that cause temporary delirium. He might have recovered more quickly, they believe, if Willis's medical treatments had not so weakened him that they "aggravated the underlying condition."[23] But in 1789, the return of the king's sanity was, for the mad-doctors, a medical triumph of the most visible sort.

In the wake of the king's recovery, a number of English physicians raced to exploit the commercial opportunity at hand by publishing their novel methods for curing insanity. Their marketing message was often neat as a twentieth century sound bite: "Insanity proved curable."[24] One operator of a madhouse in Chelsea, Benjamin Faulkner, even offered a money-back guarantee: Unless patients were cured within six months, all board, lodging, and medical treatments would be provided "free of all expence whatever."[25] The mad trade in England flourished. The number of private madhouses in the London area increased from twenty-two in 1788 to double that number by 1820, growth so stunning that many began to worry that insanity was a malady particularly common to the English.

In this era of medical optimism, English physicians—and their counterparts in other European countries—developed an ever more innovative array of therapeutics. Dunking the patient in water became quite popular—a therapy intended both to cool the patient's scalp and to provoke terror. Physicians advised pouring buckets of water on the patient from a great height or placing the patient under a waterfall; they also devised machines and pumps that could pummel the patient with a torrent of water. The painful blasts of water were effective "as a remedy and a punishment," one that made patients "complain of pain as if the lateral lobes of the cerebrum were split asunder."[26] The Bath of Surprise became a staple of many asylums: The lunatic, often while being led blindfolded across a room, would suddenly be dropped through a trapdoor into a tub of cold water—the unexpected plunge hopefully inducing such terror that the patient's senses might be dramatically restored. Cullen found this approach particularly valuable:

> Maniacs have often been relieved, and sometimes entirely cured, by the use of cold bathing, especially when administered in a certain manner. This seems to consist, in throwing the madman in the cold water by surprise; by detaining him in it for some length of time; and pouring water frequently upon the head, while the whole of

> the body except the head is immersed in the water; and thus managing the whole process, so as that, with the assistance of some fear, a refrigerant effect may be produced. This, I can affirm, has been often useful.[27]

The most extreme form of water therapy involved temporarily drowning the patient. This practice had its roots in a recommendation made by the renowned clinician of Leyden, Hermann Boerhaave. "The greatest remedy for [mania] is to throw the Patient unwarily into the Sea, and to keep him under Water as long as he can possibly bear without being quite stifled."[28] Burrows, reviewing this practice in 1828, said it was designed to create "the effect of asphyxia, or suspension of vital as well as of all intellectual operations, so far as safety would permit."[29] Boerhaave's advice led mad-doctors to concoct various methods for simulating drowning, such as placing the patient into a box drilled with holes and then submerging it underwater. Joseph Guislain built an elaborate mechanism for drowning the patient, which he called "The Chinese Temple." The maniac would be locked into an iron cage that would be mechanically lowered, much in the manner of an elevator car, into a pond. "To expose the madman to the action of this device," Guislain explained, "he is led into the interior of this cage: one servant shuts the door from the outside while the other releases a break which, by this maneuver, causes the patient to sink down, shut up in the cage, under the water. Having produced the desired effect, one raises the machine again."[30]

The most common mechanical device to be employed in European asylums during this period was a swinging chair. Invented by Englishman Joseph Mason Cox, the chair could, in one fell swoop, physically weaken the patient, inflict great pain, and invoke terror—all effects perceived as therapeutic for the mad. The chair, hung from a wooden frame, would be rotated rapidly by an operator to induce in the patient "fatigue, exhaustion, pallor, horripilatio [goose bumps], vertigo, etc," thereby producing "new associations and trains of thoughts."[31] In the hands of a skilled operator, able to rapidly alter the directional motion of the swing, it could reliably produce nausea, vomiting, and violent convulsions. Patients would also involuntarily urinate and defecate, and plead for the machine to be stopped. The treatment was so powerful, said one nineteenth-century physician, that if the swing didn't make a mad person obedient, nothing would.[32]

Once Cox's swing had been introduced, asylum doctors tried many variations on the theme—spinning beds, spinning stools, and spinning boards were all introduced. In this spirit of innovation and medical advance, one inventor built a swing that could twirl four patients at once, at revolutions up to 100 per minute. Cox's swing and other twirling devices, however, were eventually banned by several European governments, the protective laws spurred by a public repulsed by the apparent cruelty of such therapeutics. This governmental intrusion into medical affairs caused Burrows, a madhouse owner who claimed that he cured 91 percent of his patients, to complain that an ignorant public would "instruct us that patient endurance and kindliness of heart are the only effectual remedies for insanity!"[33]

Even the more mainstream treatments—the Bath of Surprise, the swinging chair, the painful blistering—might have given a compassionate physician like Rush pause. But mad-doctors were advised to not let their sentiments keep them from doing their duty. It was the highest form of "cruelty," one eighteenth-century physician advised, "not to be bold in the Administration of Medicine."[34] Even those who urged that the insane, in general, should be treated with kindness, saw a need for such heroic treatments to knock down mania. "Certain cases of mania seem to require a boldness of practice, which a young physician of sensibility may feel a reluctance to adopt," wrote Thomas Percival, setting forth ethical guidelines for physicians. "On such occasions he must not yield to timidity, but fortify his mind by the councils of his more experienced brethren of the faculty."[35]

Psychiatry in America

It was with those teachings in mind that Rush introduced medical treatments into the regimen of care at Pennsylvania Hospital. Although he was a Quaker, a reformist, and one who could empathize with the unfortunate, he was also an educated man, confident in the powers of science, and that meant embracing the practices advocated in Europe. "My first principles in medicine were derived from Dr. Boerhaave," he wrote, citing as his inspiration the very physician who had dreamed up drowning therapy.[36] Moreover, at the time, he and other leading American doctors were struggling to develop an academic foundation for their profession, with European medicine the model to emulate. Before the American Revolution, fewer than 5 percent of the 3,500 doctors in the country had degrees, and only about 10 percent had any formal training at all. Medicine in colonial America had a well-deserved reputation as a refuge for quacks. But that was changing. In 1765, the first medical school in America had been established at the College of Philadelphia, where Rush was one of the faculty members. In the 1790s, medical societies were formed, and the first periodical medical journal was published. It all led to a proud sense of achievement—American medicine was now a scientific discipline. "There were the usual comments that more had been achieved in science over the preceding hundred years than in all the past centuries," wrote historian Richard Shryock. "Now and then, [there was] even a hint that there was little left for posterity to do in the medical line."[37]

Rush's conception of madness reflected the teachings of his European mentors. He believed that madness was caused by "morbid and irregular" actions in the blood vessels of the brain.[38] This abnormal circulation of the blood, he wrote, could be due to any number of physical or psychological causes. An injury to the brain, too much labor, extreme weather, worms, consumption, constipation, masturbation, intense study, and too much imagination could all cause a circulatory imbalance. To fix this circulatory disorder, he advocated the copious bleeding of patients, particularly those with mania. He drew 200 ounces of blood from one patient in less than two months; in another instance, he bled a manic patient forty-seven times, removing nearly four gallons of blood. As much as "four-fifths of the blood in the body" should be drawn away, he said. His bleeding regimen was so extreme that other doctors publicly criticized it as a "murderous dose" and a "dose for a horse," barbs that Rush dismissed as the talk of physicians competing "for business and money."[39]

As he employed other remedies he'd learned from the Europeans, he did so in ways that fit his belief that madness was due to a circulatory disorder. For instance, he argued that blisters should be raised on the ankles rather than the scalp, as this would draw blood away from the overheated head. Caustics could be applied to the back of the neck, the wound kept open for months or even years, as this would induce a "permanent discharge" from the overheated brain. The head could also be directly treated. The scalp could be shaved and cold water and ice dumped on the overheated brain. Purges and emetics could also draw blood away from the inflamed brain to the stomach and other organs. Rush administered all of these treatments confident that they were scientific and worked by helping to normalize blood flow in the brain.

Although Rush constantly preached the need to treat the insane in a kind manner, at times he adopted the language of his English teachers, comparing lunatics to the "tyger, the mad bull, and the enraged dog." Intimidation tactics could be used to control them; patients might even be threatened with death. "Fear," he said, "accompanied with pain and a sense of shame, has sometimes cured this disease." A doctor in Georgia, he recounted, had successfully cured a madman by dropping him into a well, the lunatic nearly drowning before he was taken out. Concluded Rush: "Terror acts powerfully upon the body, through the medium of the mind, and should be employed in the cure of madness."[40]

Rush also made use of spinning therapy. Patients suffering from melancholy, or "torpid madness," would be strapped horizontally to a board that could be mechanically spun at great speeds, a device he called the gyrator. He reasoned this version of madness was caused by too little blood circulation in the head (rather than the fullness of circulation that led to mania) and that by placing the patient with his or her feet at the board's fixed point of motion, blood would rush to the brain. The treatment also made the mad so weak and dizzy that any wild thoughts would be temporarily driven from the brain. Burrows, who urged that every modern asylum should have a gyrator in its medical arsenal, said that it could instill fear in even the most hopeless cases.

> Where no expectation of cure has been entertained, a few trials have produced a wonderful improvement in manners and behaviour. Where the degree of violence has been so great as to compel a

> rigid confinement, the patient has become tractable, and even kind and gentle, from its operation. The morbid association of ideas has been interrupted, and even the spell of the monomaniac's cherished delusion broken.[41]

Rush was particularly proud of the "Tranquilizer Chair" he invented, which he boasted could "assist in curing madness." Once strapped into the chair, lunatics could not move at all—their arms were bound, their wrists immobilized, their feet clamped together—and their sight was blocked by a wooden contraption confining the head. A bucket was placed beneath the seat for defecation, as patients would be restrained for long periods at a time. Rush wrote:

> It binds and confines every part of the body. By keeping the trunk erect, it lessens the impetus of blood toward the brain. By preventing the muscles from acting, it prevents the force and frequency of the pulse, and by the position of the head and feet favors the easy application of cold water or ice to the former and warm water to the latter. Its effects have been truly delightful to me. It acts as a sedative to the tongue and temper as well as to the blood vessels. In 24, 12, six and in some cases in four hours, the most refractory patients have been composed. I call it a Tranquilizer.[42]

This was the first American therapeutic for insanity that was exported back to the Europeans. Asylum physicians eagerly embraced it, finding that it would "make the most stubborn and irascible patients gentle and submissive," and since patients found it painful, "the new and unpleasant situation engages his attention and directs it toward something external."[43] One told of keeping a patient in the chair for six months.

Rush stood at the very pinnacle of American medicine at that time. He was the young country's leading authority on madness, and other American physicians copied his ways. They too would bleed their insane patients and weaken them with purges, emetics, and nausea-inducing drugs. Physicians familiar with his teachings might also use water therapies. A Delaware physician, writing in an

1802 medical journal, told of the dousing therapy he'd utilized while treating an insane man confined at home. "He was chained to the floor, with his hands tied across his breast—clothes torn off, except the shirt—his feet and elbows bruised considerably—and his countenance, grimaces and incoherent language, truly descriptive of his unhappy condition. As he was free from fever, and his pulse not tense or preternaturally full, I deemed his a fair case for the application of cold water."[44]

At least a few early American physicians tested the merits of drowning therapy. A Dr. Willard, who ran a private madhouse in a small town near the border of Massachusetts and Rhode Island, used this European technique as part of his efforts "to break the patient's will and make him learn that he had a master." Dr. Willard's methods were carefully described by Isaac Ray, a prominent nineteenth-century psychiatrist:

> The idea was . . . that if the patient was nearly drowned and then brought to life, he would take a fresh start, leaving his disease behind. Dr. Willard had a tank prepared on the premises, into which the patient, enclosed in a coffin-like box with holes, was lowered by means of a well-sweep. He was kept there until bubbles of air cease to rise, then was taken out, rubbed and revived.[45]

There don't appear to be any historical accounts from patients recording what it was like to endure this therapy. But a history of Brattleboro, Vermont, written in 1880, does describe briefly the reaction of Richard Whitney—a prominent Vermont citizen—to being plunged, one day in 1815, headfirst into the water and held there until all air had left his lungs:

> A council of physicians . . . decided upon trying, for the recovery of Mr. Whitney, a temporary suspension of his consciousness by keeping him completely immersed in water three or four minutes, or until he became insensible, and then resuscitate or awaken him to a new life. Passing through this desperate ordeal, it was hoped, would divert his mind, break the chain of unhappy associations, and thus remove the cause of his disease. Upon trial, this system of regeneration proved of

> no avail for, with the returning consciousness of the patient, came the knell of departed hopes, as he exclaimed, "You can't drown love!"[46]

The Vermont physicians, thus disappointed, turned to opium as a cure, a treatment that subsequently killed the lovesick Richard Whitney.

2

THE HEALING HAND OF KINDNESS

If there is any secret in the management of the insane, it is this: respect them and they will respect themselves; treat them as reasonable beings, and they will take every possible pain to show you that they are such; give them your confidence, and they will rightly appreciate it, and rarely abuse it.

—Samuel Woodward[1]

IN 1812, BENJAMIN RUSH collected his thoughts on madness in a book, *Medical Inquiries and Observations Upon the Diseases of the Mind.* It was the first psychiatric text to be published in the United States, and Rush had every reason to believe that his counsel would guide American physicians for decades to come. He had summarized the medical teachings of elite European doctors with his own variations on the theme, and he had provided readers with a reasoned explanation as to why the various medical treatments could cure the mad. Yet in a very short time, his gyrator would be banished from Pennsylvania Hospital, perceived as an instrument of abuse, and even his prized tranquilizer chair would come to be seen as an embarrassing relic from an unenlightened past.

The reason was the rise of moral treatment.

Rush, in his writings and in his hospital practices, had actually synthesized two disparate influences from Europe. The medical treatments he advised—the bleedings, the blisterings, the psychological terror—were the stuff of medical science. His counsel that the mentally ill should be treated with great kindness reflected reformist practices, known as moral treatment, that had arisen in France and among Quakers in England. However, the two influences made for strange bookfellows, for moral treatment had come about, in large part, in response to the harsh medical therapeutics. Care of the mentally ill was at a crossroads when Rush died in 1813, and it was moral treatment that took hold in the next few years, remaining the therapeutic ideal for most of the nineteenth century.

Lunacy Reform in Europe

The seeds of moral treatment were planted in 1793, when the French Revolution was raging, and physician Philippe Pinel was appointed by the revolutionary government to tend to the insane at the Salpêtrière and Bicêtre asylums in Paris. Prior to the revolution, when France was ruled by King Louis XVI, the lunatics had been treated with the usual neglect and brutality. Those who were manic were kept in chains and fetters, and all suffered the extremes of heat and cold in their miserable cells. At Bicêtre, which was the asylum for men, the insane were fed only one pound of bread a day, which was doled out in the morning, leaving them to spend the remainder of the day "in a delirium of hunger."[2] More than half of the men admitted to the asylums died within a year from starvation, cold, and disease. But the rallying cry of the French Revolution was *liberté, égalité, fraternité,* and by the time Pinel arrived, a lay superintendent, Jean Baptiste Pussin, had begun to treat them better. Pussin increased the patients' rations and reduced the use of restraints. Pinel, who greatly admired Pussin, quickly noticed that if the insane were not treated cruelly, they behaved in a fairly orderly fashion. The rantings and ravings that appeared to define the mad—the tearing of clothes, the smearing of feces, the screaming—were primarily antics of protest over inhumane treatment.

> I saw a great number of maniacs assembled together, and submitted to a regular system of discipline. Their disorders presented an endless variety of character; but their discordant movements were regulated on the part of the governor [Pussin] by the greatest possible skill, and even extravagance and disorder were marshalled into order and harmony. I then discovered, that insanity was curable in many instances, by mildness of treatment and attention to the state of the mind exclusively, and when coercion was indispensable, that it might be very effectively applied without corporal indignity.[3]

Inspired by Pussin, Pinel set out to rethink care of the insane. He took his patients' case histories and carefully detailed their responses to the treatment they received. He was highly skeptical about the remedies prescribed in medical texts and found that they did little to help his patients. The treatments were, he concluded, "rarely useful and frequently injurious" methods that had arisen from "prejudices, hypotheses, pedantry, ignorance, and the authority of celebrated names." Recommendations that the blood of maniacs be "lavishly spilled" made him wonder "whether the patient or his physician has the best claim to the appellation of a madman." His faith in "pharmaceutic preparations" declined to the point that he decided "never to have recourse to them," except as a last resort.

In place of such physical remedies, Pinel decided to focus on the "management of the mind," which he called *"traitement morale."* He talked to his patients and listened to their complaints. As he got to know them, he came to appreciate their many virtues. "I have nowhere met, except in romances, with fonder husbands, more affectionate parents, more impassioned lovers, more pure and exalted patriots, than in the lunatic asylum, during their intervals of calmness and reason. A man of sensibility may go there every day of his life, and witness scenes of indescribable tenderness to a most estimable virtue."

The success of this approach—not only did patients behave in a more orderly fashion, but some began talking sufficient sense to be discharged—convinced Pinel that prevailing scientific notions about the causes of insanity were wrong. If a nurturing environment could heal, he reasoned in his 1801 treatise, *Traité médico-philosophique sur l'aliénation mentale,* then insanity was not likely due to an "organic lesion of the brain." Instead, he believed that many of his patients had retreated into delusions or become overwhelmed with depression because of the shocks of life—disappointed love, business failures, the blows of poverty.

In his treatise, Pinel set forth a vision for building a therapeutic asylum for the insane. Physicians would be schooled in distinguishing among the different types of insanity (he identified five "species" of mental derangement), and patients would be treated with therapies suitable for their particular kind of madness. The hospital, meanwhile, would be organized so that the patients' time would be filled not with idleness but with activities—work, games, and other diversions. Attendants would be counseled to treat the patients with "a mildness of tone" and never to strike them. A lay superintendent, imbued with a humanitarian philanthropy toward the mentally ill, would govern the asylum. In such a hospital, Pinel said, "the resources of nature" could be "skillfully assisted in her efforts" to heal the wounded mind.

As dramatic as Pinel's reform ideas were, they were still those of a medical man—he was seeking to change how physicians and society treated the insane but was not questioning whether the insane should be placed under the care of doctors. During this same period, Quakers in York, England, were developing their own form of moral treatment, and their reform efforts presented a much more vigorous challenge to the medical establishment. And while Pinel is remembered as the father of moral treatment, it was the Quakers' reforms, rooted in religious beliefs, that most directly remade care of the insane in America.

In eighteenth-century England, Quakers were largely shunned as outcasts. The Quaker movement had been founded in the 1650s by people dissatisfied with the authoritarian and class-conscious ways of the Protestant Church, and they were a socially radical group. They refused to pay tithes to the church, bear arms, or show obeisance to the king. They chose to live as a "Society of Friends" in a simple and plain manner, adopted pacifism as a guiding tenet, and believed that all people were equal before God, each soul guided by an "inner light." Although the Quakers were often persecuted for their beliefs—they were not allowed, for example, to earn degrees from the two universities in England—they prospered as merchants and farmers, and this commercial success strengthened their confidence and resolve to keep their distance from the ruling elite. They viewed doctors with a great deal of skepticism and mistrust, and their mistrust grew after one of their members, a young woman named Hannah Mills, died in 1791 of ill treatment and neglect at the York asylum.

The York Quakers made no noisy protest about her death. That was not their way. Instead, led by William Tuke, they quietly decided to build their own "retreat" for their mentally ill, one that would be governed by their religious values rather than by any professed medical wisdom. They would treat the ill with gentleness and respect, as the "brethren" they were. It would be the needs of the ill, and not the needs of those who managed the retreat, that would guide their care.

The Quakers opened their small home in 1796. It was a simple place, with gardens and walks where the ill could get their fill of fresh air. They fed patients four times daily and regularly provided snacks that included biscuits along "with a glass of wine or porter."[4] They held tea parties, at which the patients were encouraged to dress up. During the day, patients were kept busy with a variety of tasks—sewing, gardening, and other domestic activities—and given opportunities to read, write, and play games like chess. Poetry was seen as particularly therapeutic.

The Quakers borrowed their "medical" philosophy from the ancient wisdom of Aeschylus: "Soft speech is to distemper'd wrath, medicinal." The therapeutics of the English mad-doctors, wrote Samuel Tuke, William's grandson, in 1813, were those at which "humanity should shudder." The one medical remedy regularly employed at the York Retreat was a warm bath, which was to last from twenty minutes to an hour. "If it be true," Samuel Tuke reasoned, "that oppression makes a *wise* man mad, is it to be supposed that stripes, and insults, and injuries, for which the receiver knows no cause, are calculated to make a *madman* wise? Or would they not exasperate his disease, and excite his resentment? May we not hence most clearly perceive, why furious mania, is almost a stranger in the Retreat? Why all patients wear clothes, and are generally induced to adopt orderly habits?"

In this gentle environment, few needed to be confined. There was rarely a day when as many as two patients had to be secluded at the same time—seclusion in a dark, quiet room being the common practice for controlling rowdy patients. In its first fifteen years of operation, not a single attendant at the York Retreat was seriously injured by a violent patient. Nor was this cooperative behavior the result of a patient group that was only mildly ill—the majority had been "insane" for more than a year, and many had been previously locked up in other English asylums, where they were viewed as incurable.

The Quakers, humble in nature, did not believe that their care would unfailingly help people recover. Many would never get well, but they could still appreciate living in a gentler world and could even find happiness in such an environment. As for the path to true recovery, the Quakers professed "to do little more than assist nature." They wouldn't even try to talk their patients out of their mad thoughts. Rather, they would simply try to turn their minds to other topics, often engaging them in conversation about subjects their patients were well versed in. In essence, the Quakers sought to hold up to their patients a mirror that reflected an image not of a wild beast but of a worthy person capable of self-governance. "So much advantage has been found in this institution from treating the patient as much in the manner of a rational being, as the state of his mind will possibly allow," Tuke said.

Their simple, common-sense methods produced good results. During the York Retreat's first fifteen years, 70 percent of the patients who had been ill for less than twelve months recovered, which was defined by Tuke as never relapsing into illness. Even 25 percent of the patients who had been chronically ill before coming to the retreat, viewed as incurable, recovered under this treatment and had not relapsed by 1813, the year Tuke published *Description of the Retreat.*

Moral Treatment in America

Together, Pinel and the York Quakers had presented European society (and by extension American society) with a new way to think about the mad. No longer were they to be viewed as animals, as creatures apart. They were, instead, to be seen as beings within the human family—distressed people to be sure, but "brethren." The mad had an inner capacity for regaining self-control, for recovering their reason. The ultimate source of their recovery lay inside themselves, and not in the external powers of medicine.

This was a radical change in thinking, yet in the early 1800s, it was a belief that American society was primed to embrace. It fit the democratic ideals that were so fresh in the American mind, and the optimistic tenor of the times. The class distinctions so prevalent in the 1700s had given way to a belief that in democratic America, the common man could rise in status. Many preachers, as historian Gerald Grob has noted, stopped threatening their flocks with eternal damnation and instead delivered uplifting sermons about how people could enjoy God's grace while on this good Earth. Personal transformation was possible. Moreover, the good society was one that would, in the words of Andrew Jackson, "perfect its institutions" and thereby "elevate our people."[5] And what group was more in need of transformation, of being touched by God's grace, and of being "elevated," than the beleaguered souls who'd lost their reason?

Philadelphia Quakers opened the first moral-treatment asylum in America in 1817, and soon others appeared as well. The social elite of Boston, led by members of the Congregational Church, established one in 1818, which later became known as McLean Hospital. Bloomingdale Asylum in New York City opened in 1821, on the site of what is now Columbia University, a project that was guided by Quaker Thomas Eddy. Three years later, the Hartford Retreat in Connecticut began accepting patients. All of these asylums were privately funded, primarily catering to well-to-do families, but soon states began building moral-treatment asylums for the insane poor. The first such public asylum opened in Worcester, Massachusetts, in 1833, and by 1841, there were sixteen private and public asylums in the United States that promised to provide moral treatment to the insane.

The blueprint for a moral-treatment asylum was fairly sharply defined. The facility was to be kept small, providing care to no more than 250 patients. It should be located in the country, the grounds graced by flowerbeds and gardens, where the mentally ill could take their fill of fresh air and find solace in tending to plants. The building itself should be architecturally pleasing, even grand in its nature—the insane were said to be particularly sensitive to aesthetic influences. Most important, the asylum was to be governed by a superintendent who was "reasonable, humane . . . possessing stability and dignity of character, mild and gentle . . . compassionate."[6] He would be expected to know his patients well, eat with them, and, in the manner of a father figure, guide them toward a path of reason.

Each day, a variety of activities would keep patients busy, which it was hoped would divert their thoughts from their obsessions and paranoid ideas. They would spend their time gardening, reading, playing games, and enjoying educational pursuits. Theater groups would be invited in to perform; speakers would give after-dinner talks. In this environment, restraints were to be used as a last resort. Instead, a ward system that rewarded good behavior would keep patients in line. Those who were disruptive would be placed on ground floors furthest from the social center of the asylum. Those who behaved would get the preferred rooms on the top floors, and they would also be given extra liberties. They would be allowed to stroll about the grounds and be given the privilege of going into town, as long as they pledged not to drink and to return to the asylum on time.

By treating the mentally ill in this manner, it was hoped that they would regain the ability to control their behavior and their thoughts, and through the application of their will, maintain their recovery even after discharge. The basic therapeutic principle, said Dr. Eli Todd, superintendent at the Hartford Retreat, was "to treat [the insane] in all cases, as far as possible, as rational beings."[7]

To a remarkable degree, the asylums followed this blueprint during their initial years. Visitors, who included Charles Dickens, regularly came away impressed. Patients at McLean Hospital often spent their days rowing on the Charles River or taking carriage rides. Patients formed baseball teams, published their own newspapers, and attended nightly lectures. They were allowed to eat with knives and forks, and few problems resulted from their being trusted in this manner. They would hold their own meetings and pass resolutions for self-governance, setting down expectations for proper behavior by their peers. Asylums regularly held lunatic balls, at which visitors, although noticing that the patients might dance strangely, would wonder where all the real lunatics were. A reporter for *Harper's* magazine summed up the experience: "The asylum on Blackwell's Island [in New York City] is, throughout, perfect in respect of cleanliness, order and comfort."[8]

Moral treatment appeared to produce remarkably good results. Hartford Retreat announced that twenty-one of twenty-three patients admitted in its first three years recovered with this gentle treatment. At McLean Hospital, 59 percent of the 732 patients admitted between 1818 and 1830 were discharged as "recovered," "much improved," or "improved." Similarly, 60 percent of the 1,841 patients admitted at Bloomingdale Asylum in New York between 1821 and 1844 were discharged as either "cured" or "improved." Friends Asylum in Philadelphia regularly reported that approximately 50 percent of all admissions left cured. Even the state hospitals initially reported good outcomes. During Worcester State Lunatic Asylum's first seven years, more than 80 percent of those who had been ill for less than a year prior to admission "recovered," which meant that they could return to their families and be expected to function at an acceptable level.[9]

All of this created a sense of great optimism, a belief that in most instances, insanity could be successfully treated. "I think it is not too much to assume that insanity," wrote Worcester superintendent Samuel Woodward in 1843, "is more curable than any other disease of equal severity; more likely to be cured than intermittent fever, pneumonia, or rheumatism."[10]

Medicine's Grab of Moral Treatment

During this initial heady period, moral treatment in America did begin to stray from its Quaker roots in one significant way. In the very beginning, the asylums were run by people who shared the mistrust of the York Quakers toward mad-doctors and their medical treatments. Friends Asylum, Bloomingdale Asylum, and Boston's asylum were all governed by lay superintendents or by a physician who thought little of physical (or "somatic") remedies for madness.[11] Rush's prescribed remedies were seen as useless—or worse.

One asylum director confided that "he and his colleagues were so prejudiced against the use of medical measures, as to object even to the election of physicians in their board, being fearful they might effect some innovation."[12] Rufus Wyman, the physician superintendent at McLean Hospital, dismissed traditional medical remedies as "usually injurious and frequently fatal."[13]

However, the rise of moral treatment in the 1810s had presented physicians with a clear threat. It was evident that an age of asylum building was at hand, and yet, even as this societal response to insanity was being organized, physicians were losing out. Quakers in Philadelphia had built an asylum, and so had civic groups in Boston and New York City, and groups in other cities were sure to do the same—yet what was the physician's role in this asylum care? Marginal, at best. The Connecticut State Medical Society, sizing up this threat, rushed to beat local religious groups and the social elite to the punch. It lobbied the state and civic groups for the finances to build a local asylum, and in return for its organizational efforts, the society extracted the right to appoint the superintendent, a governance clause that insured it would be led by a physician.

When the Hartford Retreat opened in 1824, superintendent Dr. Eli Todd immediately noted that care at this asylum, even though it might be named after the York Retreat, would be different. Here, both physical remedies and moral treatment would be used to provide superior care to the insane. The Quakers in York, he said, "have placed too little reliance upon the efficacy of medicine in the treatment of insanity, and hence their success is not equal to that of other asylums in which medicines are more freely employed." The first moral-treatment asylums in the United States, he added, having modeled their efforts on the York Retreat, had repeated the mistake, resulting in treatment that "is feeble compared to the lofty conceptions of truly combined medical and moral management."[14]

Moral treatment in America had taken a decided turn. Although the reform had begun partly as a backlash against medical practices, medicine was now reclaiming it as its own. Physicians were best suited to run the new facilities. As Massachusetts, New York, and other states funded their public asylums, they accepted this argument and appointed doctors as superintendents. Asylum medicine became its own specialty, and in 1844, superintendents at thirteen asylums formed the Association of Medical Superintendents of American Institutions for the Insane (AMSAII) to promote their interests. One of AMSAII's first orders of business was to pass a resolution stating that an asylum should always have a physician as its chief executive officer and superintendent.

As promised by Todd, asylum physicians injected medical remedies into the moral-treatment regimen. They used mild cathartics, bloodletting on occasion, and various drugs—most notably morphine and opium—to sedate patients. Their use of such chemical "restraints," in turn, made them more receptive to the use of physical restraints, which they increasingly turned to as their asylums became more crowded. Mitts and straitjackets eventually became commonplace. Utica Lunatic Asylum in New York devised a crib with a hinged lid for confining disruptive patients at night, a space so claustrophobic that patients would fight violently to get free, before finally collapsing in exhaustion. In 1844, AMSAII formally embraced the use of physical restraints, arguing that to completely forgo their use "is not sanctioned by the true interests of the insane."[15]

As physicians gained control of the asylums, they also constructed a new explanation for the success of moral treatment—one that put it back into the realm of a physical disorder. Pinel's "non-organic" theory would not do. If it were not a physical ailment, then doctors would not have a special claim for treating the insane. Organizing activities, treating people with kindness, drawing warm baths for the ill—these were not tasks that required the special skills of a physician. Although individual doctors had their pet theories, a consensus arose that mental disorders resulted from irritated or worn-out nerves. The exhausted nerves transmitted faulty impulses to the brain (or from one brain region to another), and this led to the hallucinations and odd behavior characterized by madness. Moral treatment worked as a *medical* remedy precisely because it restored, or otherwise soothed, the irritated nerves. The pastoral environment, the recreational activities, and the warm bath were all medical tonics for the nervous system.

This conception of madness, of course, was quite at odds with Rush's. He had theorized that madness was caused by a circulatory disorder—too much blood flowing to the head. But the asylum doctors were loath to admit that the medical texts of the past had been in error. It was difficult to believe, wrote Earle Pliny, superintendent at Bloomingdale Asylum, that Rush, "an acute and sagacious observer, a learned and profound medical philosopher" had been wrong.[16] Moral treatment now worked, they explained, because the physical causes of madness had changed.

In Rush's time, Pliny and others reasoned, people had lived closer to vigorous nature and thus were likely to fall ill because of a surplus of strength and energy. Such a disorder, Pliny said, "required a more heroic method of attack for its subjection." But in the nineteenth century, people no longer lived so close to nature. Instead, their nerves could be worn down by the demands of civilization. The striving for success, the financial pressures, and the opportunities that democratic societies and capitalism offered—all were sources of mental illness. "Insanity," declared Edward Jarvis, a physician who researched asylum care, "is part of the price which we pay for civilization."[17]

It was, in its own way, an artful construction. In the eighteenth century, medicine and science had developed an armamentarium of harsh treatments—the bleedings, the blisterings, the psychological terror, the spinning devices, the starvation diets—because the mad were closer in nature to wild animals and thus in need of therapies that would deplete their energy and strength. But now the insane and mentally ill were worn down by the travails of modern society and thus no longer needed such harsh remedies. Instead, they required the nurturing care of moral treatment. Medicine, with all its agility and wisdom, had simply developed new therapies for a changed disease.

Moral Treatment at Its Best

The forces that would lead to the downfall of moral treatment began appearing in the 1840s, and before the end of the century, it would be disparaged as a hopelessly naive notion, a form of care that had never produced the positive results initially claimed by the asylum doctors. Yet it was during this period of downfall that moral treatment, in a form that would remind the future of its potential to heal, was best practiced. For more than forty years, from 1841 to 1883, moral treatment held sway at the Pennsylvania Hospital for the Insane, during which time the asylum was continually governed by a memorable Quaker physician, Thomas Kirkbride.[18]

Kirkbride, born on July 31, 1809, was raised on a 150-acre farm in Pennsylvania. His family faithfully observed Quaker religious traditions, and as a child he attended religious schools run by the Friends Society. This upbringing shaped his adult character: He was humble, soft-spoken, simple in his dress, and reflective in his thoughts. His faith fostered a confident belief that all people could amend their ways. After he graduated from the University of Pennsylvania, he did his residency at Friends Asylum in Frankford, and it was there that he soaked up the principles of moral treatment in a form still close to its Quaker roots.

The new asylum that Pennsylvania Hospital opened in 1841, in the countryside west of Philadelphia, was an opulent place. It had a lovely dining room, a day room for playing games, and even a bowling alley. Kirkbride added a greenhouse and museum, complete with stuffed birds, for the patients' amusement. Flowerbeds and meticulous landscaping furthered the sense of pastoral comfort. "It should never be forgotten," Kirkbride happily wrote, "that every object of interest that is placed in or about a hospital for the insane, that even every tree that buds, or every flower that blooms, may contribute in its small measure to excite a new train of thought, and perhaps be the first step towards bringing back to reason, the morbid wanders of the disordered mind."[19]

Kirkbride embraced all the usual methods of moral treatment, applying them with unflagging energy. Patients, roused from their beds at 6 A.M. sharp, exercised daily in the gymnasium. They often dressed well, men in suits and ties and women in fine dresses, and during the afternoon they would pleasantly pass hours in the reading parlor, which had 1,100 volumes. Teachers were hired to give classes in reading and sewing. Evening entertainment at the asylum featured magic-lantern shows, guest lectures, concerts, and theatrical performances, a parade of activities that became famous locally for their high quality. At night, patients retired to semiprivate rooms that could have done a modest hotel proud. The chest of drawers, mirror, and wall paintings in each room helped patients feel respected and surrounded by comfort.

Kirkbride made a special effort to hire attendants who had the temperament to treat the patients well. They were not to consider themselves as "keepers" of the insane but rather as their "attendants" and companions. He required all job applicants to provide references attesting to their good character and sought only to employ those who had "a pleasant expression of face, gentleness of tone, speech and manner, a fair amount of mental cultivation, imperturbable good temper, patience under the most trying provocation, coolness and courage in times of danger, cheerfulness without frivolity."[20] Attendants were given rule books and knew that they would be dismissed if they hit a patient.

It all led, as the visitor Dr. George Wood reported in 1851, to a hospital graced with decorum and seeming tranquillity.

> Scattered about the ground, in the different apartments of the main building, or in the out-houses, you encounter persons walking, conversing, reading or variously occupied, neatly and often handsomely dressed, to whom as you pass you receive an introduction as in ordinary social life; and you find yourself not unfrequently quite at a loss to determine whether the persons met with are really the insane, or whether they may not be visitors or officials in the establishment.[21]

However, what most distinguished the care at the hospital was Kirkbride's skill as a healer. At this asylum, the doctor-patient relationship was *the* critical element in the curative process. In his counseling of patients, Kirkbride would gently encourage them to develop friendships, dress well, and rethink their behavior. They would need to stop blaming their families for having committed them and acknowledge instead that they had behaved poorly toward their families and needed to reestablish ties with them. Developing a sense of guilt and even shame for one's misbehavior—a social conscience, in other words—was part of acquiring a new perception of one's self. Most important of all, he preached to his patients that they could, through their exercise of their free will, choose to be sane. They could resist mad thoughts and fight off their attacks of depression and mania. They were not hopelessly ill, they were not forever broken people, but rather they had the potential to get better and to stay better. "You have it almost entirely in your power to continue to enjoy these blessings," he told them. "You must be thoroughly convinced of the importance in every point, of some regular employment, and of resisting fancies that may sometimes enter your mind, but which if harbored there can only give you uneasiness and lead you into difficulty."[22]

Many patients continued to seek Kirkbride's guidance after they were discharged. A number wrote warm letters of gratitude, referring to him as "my dear friend," "my kind and patient doctor," or "my beloved physician"—words of devotion for the gentle man who had led them from despair into a world where happiness was possible. Some recalled the hospital fondly, remembering it as a "sweet quiet home." And when madness seemed to be knocking on their door once more, several told of how they would think of their good doctor and gather strength. "It was only the other night I woke in great fright," one patient wrote. "I was too frightened to call, but I suddenly thought of Dr. Kirkbride, and, as I thought, it seemed to me, that I could see him distinctly though the room was dark, and immediately I felt that peace and freedom from danger that Dr. Kirkbride always inspired." Yet another told of asserting her will, just as he had counseled: "I have great instability of nerves and temper to contend with, but knowing the necessity of self-control I try always to exercise it."[23]

Not all patients, of course, got well under Kirkbride's care. Many chafed at the behavioral controls of moral treatment. Many remained in the hospital, part of a growing caseload of chronic cases, which would become ever more problematic for the hospital. But at its most powerful, moral treatment as practiced by Kirkbride successfully led some of the very ill through a process that produced lasting inner change. As one recovered patient put it, the Pennsylvania Hospital for the Insane was "the finest place in the world to get well."

Moral Treatment's Downfall

As a social reform, moral treatment drew on the best character traits of the American people. It required compassion toward the mentally ill, and a willingness to pay for generous care to help them get well. In the 1840s and 1850s, reformer Dorothea Dix appealed to this humanitarian impulse, and states responded with a wave of asylum building. Ironically, Dix's successful lobbying was the catalyst for the downfall of moral treatment.

Dix had a personal reason for believing in this kind of care. As a young woman, she'd suffered a breakdown and, to help her recover, her family had sent her to Liverpool to live with the family of William Rathbone, whose grandfather was William Tuke. She spent more than a year there, resting at their home and becoming schooled in the reform wrought by the York Quakers. Upon returning to the United States, she vowed to bring this humane care to all of America's insane. She was a tireless and brilliant lobbyist. In state after state, she would survey the treatment of the mentally ill in local prisons and poorhouses, which inevitably turned up at least a few horror stories, and then she reported on their mistreatment to state legislatures with great literary flair. There were, she dramatically told the Massachusetts State Legislature in 1843, "Insane Persons confined in this Commonwealth in cages, closets, cellars, stalls, pens! Chained, naked, beaten with rods, and lashed into obedience!"[24] In response to her vivid appeals, twenty states built or enlarged mental hospitals. In 1840, only 2,561 mentally ill patients in the United States were being cared for in hospitals and asylums. Fifty years later, 74,000 patients were in state mental hospitals alone. The number of mental hospitals in the country, private and public, leaped from eighteen in 1840 to 139 in 1880.

However, people with all kinds of illnesses, physical as well as mental, were being put into the institutions. Syphilitics, alcoholics, and the senile elderly joined the newly insane in these hospitals, and this flood of diverse patients doomed moral treatment.

A key principle of this therapy was that it required a small facility, one that provided a homelike atmosphere. Superintendents even spoke of their patients and staff as an extended family. AMSAII argued that no asylum should ever shelter more than 250 patients.

But the rush of insane patients into state hospitals made it impossible to keep the facilities small. By 1874, state mental hospitals had on average 432 patients. One-third of the hospitals had more than 500 patients, and a few had more than 1,000. In such crowded asylums, there was little possibility that the superintendent could provide the empathy and guidance that was considered vital to helping disturbed people get well.

Moreover, from the beginning, states had been hesitant to fully duplicate the opulent ways of the private asylums. When Worcester State Lunatic Asylum was constructed, cheaper brick was used rather than stone—a small thing, but symptomatic of the cost-saving shortcuts to come. As more and more patients were sent to public asylums, states cut costs by forgoing the day rooms, the reading parlors, the bathing facilities, and the other amenities that were essential to moral treatment. Recreational activities, magic-lantern shows, and educational programs all disappeared. The insane poor were indeed now being kept in "hospitals," but they weren't receiving moral treatment as envisioned by the Quakers in York.

It all quickly snowballed. Superintendents at state asylums, where wages were pitifully low, had little hope of hiring attendants who showed "pleasantness of expression" and "softness of tone." They had to settle instead for staff drawn from the lowest rungs of society, "criminals and vagrants" in the words of one superintendent, who weren't likely to coddle the noxious patients with kindness.[25] Attendants turned to maintaining order in the old way—with coercion, brute force, and the liberal use of restraints. State legislatures, faced with soaring expenses, set up "charity boards" to oversee asylums, which quickly began to actively oppose moral treatment, with its expensive ways. Nor were the boards particularly interested in hiring devoted physicians like Kirkbride to run their asylums. Instead, they sought to hire superintendents who could manage budgets wisely and were willing to scrimp on spending for patients and, in the best manner of political appointees, grease the patronage wheels. The good superintendent was one who could ensure that supply contracts went to friends of the board.

Treatment outcomes steadily declined. During the Worcester asylum's first decade, 80 percent of its patients who had been ill less than a year before admittance were discharged as either recovered or improved. In its second decade, after the asylum was enlarged in response to Dix's appeal, this success rate dropped to 67 percent, and it continued to spiral downward in subsequent years.[26] This decline was repeated at state asylum after state asylum, which became ever more filled with chronic patients. Many of the deranged also had organic illnesses—old-age senility, cerebral arteriosclerosis, brain tumors, and dementia associated with end-stage syphilis—and thus had no hope of ever recovering. The optimism of the 1840s, when it was believed that insanity was eminently curable, turned into the pessimism of the 1870s, when it seemed that moral treatment had failed, and miserably so.*

Neurologists delivered the final blow. The Civil War, with its tremendous number of casualties, had helped produce this new medical specialty. Physicians who had become experienced in treating gunshot wounds opened private clinics after the war, touting their experience in nervous disorders. But without the war sending the wounded their way, they hungered for patients, and the crowded asylums presented an obvious solution. They needed to claim ownership of "mental disorders," and in 1878, they opened their public attack on the asylum superintendents, doing so with a haughty air of superiority.

*The obvious question today is whether moral treatment ever worked. Did treating disturbed, severely mentally ill people with kindness in small orderly retreats produce good outcomes? Modern historians have concluded that it did indeed produce surprisingly good results. In the first decades of moral treatment, 35 to 80 percent of all admitted patients were discharged within a year's time, and the majority of those discharged were viewed as having been cured. That meant that their disturbing behavior and psychotic thoughts had largely disappeared. At Pennsylvania Hospital, results remained fairly good throughout Kirkbride's tenure. Of 8,546 "insane" patients admitted from 1841 to 1882, 45 percent were discharged as cured, and another 25 percent discharged as improved. A long-term follow-up study of 984 patients discharged from Worcester asylum from 1833 to 1846, which was conducted in the 1880s, found that 58 percent had remained well throughout their lives. Another 7 percent had relapsed but had subsequently recovered and returned to the community. Only 35 percent had become chronically ill or had died while still mentally ill.[27]

As a group, the neurologists were young, confident, and aggressive. They prided themselves on being men of hard science—well schooled in anatomy and physiology, and certain that mental illness arose from lesions of the brain or nerves. They saw the asylum doctors as a pathetic lot—old, old-fashioned, and hopelessly influenced by their Christian beliefs. They were, sneered Edward Spitzka, speaking to the New York Neurological Society, little more than inept "gardeners and farmers," lousy janitors whose asylums were "moist and unhealthy," and scientific "charlatans" who knew nothing about "the diagnosis, pathology and treatment of insanity." Other leading neurologists joined in. Edward Seguin, president of the New York Neurological Society, acidly noted that one could pore through the preamble to AMSAII's constitution and "look in vain for the word science." S. Weir Mitchell, a prominent neurologist who had invented a "scientific" rest cure for mental disorders, called their treatments a fraud, their published reports incomprehensible, and asylum life "deadly to the insane." Finally, in 1879, William Hammond, who had been the nation's surgeon general during the Civil War, made their business proposition clear. Even a "general practitioner of good common sense . . . is more capable of treating successfully a case of insanity that the average asylum physician," he said. Insanity was a "brain disease" that could be successfully treated on an outpatient basis—a view of the disorder, of course, that would send patients to the neurologists' clinics.[28]

The asylum doctors didn't have much ammunition for fighting back. Most of the old guard who had pioneered moral treatment were long gone. The superintendents who had taken their place didn't have the same fire for the therapy. They were, indeed, mostly bureaucrats. Nor could the asylum doctors claim that moral therapy was a scientific therapy. Pliny and others may have fashioned a tale about how it was a medical remedy that soothed irritated nerves, but its roots were still all too clear. Moral treatment was a product of Quaker religious beliefs that love and empathy could have restorative powers. Kirkbride's genius had been in the *art* of healing rather than in any scientific understanding of the biology of madness. In 1892, the asylum superintendents officially threw in the towel and promised a new beginning. They changed the name of their association from AMSAII to the American Medico-Psychological Association and vowed to pursue scientific approaches to treating the mad. "The best definition of insanity is that it is a symptom of bodily disease," McLean superintendent Edward Cowles told his peers three years later. "Thus it is that psychiatry is shown, more than ever before, to be dependent upon general medicine."[29]

A reform that had begun a century earlier as a backlash against the harsh medical therapeutics of the day had clearly come to an end. A scientific approach to treating the mentally ill was now ready to return to the center of American medicine, and that would lead, in fairly quick fashion, to a truly dark period in American history.

PART TWO

THE DARKEST ERA

(1900–1950)

3

UNFIT TO BREED

Why do we preserve these useless and harmful beings? The abnormal prevent the development of the normal. This fact must be squarely faced. Why should society not dispose of the criminal and insane in a more economical manner?

—Dr. Alexis Carrel,
Nobel Prize winner,
Rockefeller University[1]

MORAL TREATMENT HAD represented a profound change in America's attitude toward the mentally ill. For a brief shining moment, the mentally ill were welcomed into the human family. The mad, the insane, the manic-depressive—those with mental disorders were perceived as suffering from great distress, yet still fully human. This was an attitude consonant with the noblest impulses of democracy, and with the spirit of the Declaration of Independence that "all men are created equal." Even the mad were worthy of being treated with respect and decency.

At the beginning of the twentieth century, that generous attitude toward the mentally ill disappeared in American society. It was replaced by a belief—touted as grounded in science—that the severely mentally ill were carriers of defective "germ plasm," and as such, posed a perilous threat to the future health of American society. In a stream of scientific articles, newspaper editorials, and popular books, the mentally ill were described as a degenerate strain of humanity, "social wastage" that bred at alarming rates and burdened "normal" Americans with the great expense of paying for their upkeep. America's embrace of that notion led to a wholesale societal assault on the severely mentally ill. They were prohibited from marrying in many states, forcibly committed to state hospitals in ever greater numbers, and, in a number of states, sterilized against their will. America's eugenicists even encouraged Nazi Germany in its massive sterilization of the mentally ill, a program that led directly to the crematoriums of the Holocaust.

It all began with a rather muddle-headed scientific study by Sir Francis Galton, cousin to Charles Darwin.

The Rise of Eugenics

Born in 1822, Galton enjoyed the social privileges and opportunities that come with family wealth. His family in Birmingham, England, had prospered as a maker of guns and in banking, and when his father died in 1844, young Francis inherited a sum of money that freed him from having to earn a living. He spent much of the next decade traveling through Africa, his exploration efforts in the southern part of that continent garnering a gold medal from the Royal Geographical Society. After marrying, he settled into a comfortable life in Hyde Park, hobnobbing with the elite of English society and, with time on his hands, fashioning a career as a scientist.

In 1859, when Galton was safely back in England, Darwin turned the Western world upside down with his elegant, wonderfully documented theory of evolution. Although Darwin did not specifically address humankind's beginnings in *Origin of Species*, the implication was clear: Humans had not been fashioned in one grand stroke by God but rather had evolved over time from lower animals. The agent of change in evolution was a struggle for survival, with the winners of that struggle—the fit—able to pass on their genes. In nature, the unfit were eliminated before they had an opportunity to procreate.

To Galton, this new understanding of human evolution raised an exciting possibility. If humans were not a fixed species, but one that had evolved, future change in the human makeup was not only possible but inevitable. Farmers had already demonstrated that they could breed more desirable plants and domestic animals through careful breeding practices. By applying such practices to humans, he wondered, "could not the race of men be similarly improved? Could not the undesirables be got rid of and the desirables multiplied?"[2]

Even in asking the question, Galton assumed two critical things. The first was that human society could agree on traits that were desirable. The second was that such complex traits as intelligence were intrinsic to the person rather than the result of a nurturing environment. If environment—social and educational programs—could turn out accomplished people, then society would be wise to devote its resources to improving such programs in order to improve the "race." But if intelligence and other "superior" characteristics were simply inborn, then a nation could, at least in theory, improve itself by breeding for such characteristics, much as a line of pigs might be bred for its tendency to put on weight quickly.

In 1869, Galton published a scientific work, *Hereditary Genius*, in which he concluded that it was nature, rather than nurture, that made the superior man. Galton had tracked familial relations among nearly 1,000 prominent English leaders—judges, statesmen, bankers, writers, scientists, artists, and so forth—and found that this top class came from a small, select group of people. Many were closely related. A poor person who looked at Galton's data might have decided that his study simply revealed the obvious—that in class-conscious England, privilege begat success. Galton's own life exemplified this. He had been able to make his mark as an explorer, and subsequently as a scientist, because of the wealth he had inherited. But to Galton, the data provided proof that intelligence was inherited and that a small group of successful English families enjoyed the benefits of a superior germ plasm.

Galton's notions had pronounced political implications. Humans, he had determined, were decidedly *unequal.* Democratic ideals that men were of "equal value," he said, were simply "undeniably wrong and cannot last." Even the average citizen was "too base for the everyday work of modern civilization."[3] Indeed, if a superior race were to be bred, then it would be necessary for English society—and other white societies—to encourage their fit to procreate and prevent their unfit from doing the same. Galton, for his part, imagined penning up the unfit in convents, monasteries, and asylums to prevent them from breeding. Any charity to the poor and ill, he wrote, should be conditional upon their agreeing to forgo producing offspring.

> I do not see why any insolence of caste should prevent the gifted class, when they had the power, from treating their compatriots with all kindness, so long as they maintained celibacy. But if these [compatriots] continued to procreate children inferior in moral, intellectual and physical qualities, it is easy to believe the time may come when such persons would be considered as enemies to the State, and to have forfeited all claims to kindness.[4]

In 1883, Galton coined the term "eugenics," derived from the Greek word for "well-born," as a name for the "science" that would "improve the human stock" by giving "the more suitable races or strains of blood a better chance of prevailing speedily over the less suitable than they otherwise would have had."[5] It was to be a science devoted, in large part, to dividing the human race into two classes, the eugenic and the cacogenic (or poorly born). The latter group would be tagged as having inherited bad germ plasm, and thus as a group that, at the very least, should not breed. Galton saw eugenics as a new religion, and indeed, it was a science that would have eugenicists, in essence, playing God. "What Nature does blindly, slowly, and ruthlessly, man may do providently, quickly, and kindly," he boasted.[6]

In this new eugenic view of humankind, the severely mentally ill were seen as among the most unfit. Negroes, the poor, criminals—they were all viewed as unfit to some degree. But insanity, it was argued, was the end stage of a progressive deterioration in a family's germ plasm. "Madness, when it finally breaks out, represents only the last link in the psychopathic chain of constitutional heredity, or degenerate heredity," said Austrian psychiatrist Richard von Krafft-Ebing.[7] Henry Maudsley, the most prominent English psychiatrist of his day, conceptualized insanity in similar terms. The insane patient "gets it from where his parents got it—from the insane strain of the family stock: the strain which, as the old saying was, runs in the blood, but which we prefer now to describe as a fault or flaw in the germ-plasm passing by continuity of substance from generation to generation."[8]

Although eugenics stirred much intellectual debate in England, with a few writers whipping up plays and novels on the Superman to be bred, there was little support in England, at least not before the 1920s, for eugenic laws that would prohibit the "unfit" from marrying or bearing children. But that was not the case in the United States. It was here that a society would first develop laws for compulsory sterilization of the mentally ill and other "unfit" members of society. The U.S. eugenics movement was funded by the industrial titans of America—Andrew Carnegie, John D. Rockefeller Jr., and Mary Harriman, widow of the railroad magnate Edward Harriman—and was championed, to a remarkable extent, by graduates of Harvard, Yale, and other Ivy League universities.

Eugenics in America

At the turn of the twentieth century, melting-pot America provided fertile soil for eugenics. The first great wave of immigration, in the mid-1800s, had brought more than 5 million Irish and Germans to this country. Now a second great wave of immigration was underway, with nearly 1 million immigrants arriving yearly in the first decade of the twentieth century. And this time the immigrants were even more "foreign"—Jews, Italians, Slavs. The ruling class—white Anglo-Saxon Protestants (WASPs)—saw that the United States was undergoing a great transformation, one that threatened their dominance. The country was becoming less Protestant, less English, and less white.

Not only that, the ruling class only had to look at the country's crowded slums to see which groups were breeding at the fastest rate. Once the immigrants got here, economist Francis Amasa Walker concluded in 1891, they had more children, on average, than the native born. Meanwhile, no group seemed to be less fecund than upper-class WASPs. They might have two or three children, while the Irish and their ilk kept on reproducing until their tiny walk-up apartments were filled with eight and nine children. All this resulted, the well-to-do believed, in their having to unfairly shoulder ever more costly social programs for immigrants and misfits—public schools, almshouses, and innumerable insane asylums.

The asylums were a particularly glaring example of all that was seemingly going wrong in America. In 1850, the U.S. census counted 15,610 insane in a total population of 21 million, or one out of every 1,345 people. Thirty years later, 91,997 people, in a population of 50 million, were deemed insane, or one out of every 554 people. The incidence of insanity had apparently more than doubled in thirty short years. It was a disease on the loose. And who was to blame for this frightening increase in mental illness? Although only 14 percent of the general population were immigrants, nearly 40 percent of those in state mental hospitals were foreign born.[9] Mental illness appeared to be spreading throughout the population, and from the WASP perspective, it was immigrants who were the most common carriers of this defect in germ plasm.

To the affluent, eugenics offered an explanation for what had gone wrong and a solution to the problem. In nature, the clutch of patients in the state mental asylums—along with the mentally handicapped and other misfits—would have been swiftly eliminated. But American society, with its asylums, poorhouses, and other charitable services for the weak, had—just like England—gone against nature and supported a "bad-seed" strain of humans. Any society that wanted to remain strong would do well to avoid spending on its "defectives" and would seek to keep them from breeding as well. When Andrew Carnegie read the writings of English eugenicist Herbert Spencer, who railed against social programs for the unfit, the light bulb went on for him. It was, he said, as though he had finally "found the truth of evolution."[10]

As early as 1891, American feminist Victoria Woodhull, in her book *The Rapid Multiplication of the Unfit*, argued that the "best minds" of the day agreed that "imbeciles, criminals, paupers and (the) otherwise unfit . . . must not be bred."[11] For that to occur, the unfit would have to be prohibited from marrying, segregated into asylums, and forcibly sterilized. However, that was an agenda at radical odds with democratic principles. It could only be seriously advanced if wrapped in the gauze of "neutral" science, and in 1904, Andrew Carnegie gave Harvard-educated biologist Charles Davenport the money to provide that wrapping.

Davenport, who earned his Ph.D. at Harvard and had taught zoology there, was extremely proud of his WASP heritage. He traced his ancestry back to early settlers in New England and liked to boast that he had been an American "over 300 years," for his "I" was "composed of elements that were brought to this country during the seventeenth century."[12] He was an avid reader of the writings of English eugenicists and on a trip to England dined with Galton. That excursion left him invigorated with the cause of eugenics, and upon his return, he successfully lobbied the Carnegie Foundation for funds to establish a center for the study of human evolution at Cold Spring Harbor on Long Island. Davenport received an annual salary of $3,500, making him one of the best-paid scientists in America.

Davenport approached his study of human inheritance with a Mendelian understanding of genetics. Gregor Mendel, an Austrian monk, had shown through experiments with 30,000 pea plants that inherited physical characteristics were regularly controlled by a pair of elements (or genes), with both the "male" and "female" parent (or part of the plant) contributing a gene. In plants, such physical characteristics might include size, color, and texture. In many instances, one gene type was dominant over the other. A "tall" gene for the height of a plant might be dominant over a "short" gene, and thus a combination of tall-and-short genes for height would produce a tall plant, although that plant could now pass on a short gene to its offspring. If another tall plant did the same, a short plant would result. Davenport applied this Mendelian model to complex behavioral traits in humans, each trait said to be controlled by a single gene. Moreover, he was particularly intent on proving that immigrants and societal misfits were genetically inferior, and soon he was confidently writing that people could inherit genes for "nomadism," "shiftlessness," and "insincerity." Immigrants from Italy, Greece, Hungary, and other Southeastern European countries had germ plasm that made them "more given to crimes of larceny, kidnapping, assault, murder, rape and sex-immorality." Jews inherited genes for "thieving" and "prostitution."[13]

Davenport saw a pressing need for America to act on his findings. He calculated that supporting the insane and other misfits cost taxpayers more than $100 million a year, money that was wasted because social programs had little hope of doing any good. Modern society, he complained, had "forgotten the fundamental fact that all men are created bound by their protoplasmic makeup."[14] The mentally ill and other misfits, he suggested, should not just be sterilized, but castrated. This, he said, made "the patient docile, tractable, and without sex desire."[15]

In 1910, Davenport obtained funding from Mary Harriman to establish a Eugenics Record Office at Cold Spring Harbor—an initiative that was designed to transform eugenic research findings into societal laws. Harriman, who had inherited $70 million when her husband died in 1909, donated $500,000 to the Eugenics Record Office over the next eight years. John D. Rockefeller Jr. kicked in another $22,000. Davenport used the money to gather censuslike data on the "cacogenic" in America. From 1911 to 1924, the office trained 258 field-workers, who went into mental hospitals, poorhouses, and prisons to document the family histories of the "defectives" housed there and to determine what percentage were foreign born. The field-workers also surveyed small communities, intent on identifying the percentage of misfits not yet confined by asylum walls. As a 1917 textbook, *Science of Eugenics,* approvingly explained, the Eugenics Record Office was quantifying "the burden which the unfit place upon their fellow-men."[16]

Increasingly, academics at top schools were conducting eugenic studies as well. Many of their articles were published in the *Journal of Heredity,* the house organ for the American Genetics Association. Their research typically focused on showing that the unfit were that way because of inferior genes, that they were multiplying rapidly, and that it was extremely expensive for "normals" to provide care to such "defectives." In one *Journal of Heredity* article, immigrants were likened to a "bacterial invasion." Another writer, in an article titled "The Menace of the Half Man," calculated that if the country could get rid of its defectives, then "human misery, in a well-ordered country like America, will be more than cut in half." At the same time, scholars wrung their hands over the poor job that the rich and well-born were doing at spreading their superior genes. A number of studies found that the scions of alumni of Harvard, Yale, and other Ivy League schools were a dying breed, their low birthrate a type of "race suicide." Mayflower descendants were reported, with breathless alarm, to be on their way to "extinction." And WASP women who attended elite liberal arts colleges like Wellesley were particularly deficient at having large families, leading one Ivy League academic, John Phillips, to lament that "the birth rate of college women is quite the most pathetic spectacle of all."[17]

The stream of articles signaled eugenics' arrival as an academic discipline. By 1914, forty-four colleges in America had introduced eugenics into their curriculums, with the subject taught as a science, much like engineering or mathematics, at such schools as MIT, Harvard, Columbia, Cornell, and Brown. By 1924, more than 9,000 papers on eugenics had been published, and in 1928, *Eugenical News*—a monthly newsletter published by the Eugenics Record Office—could count 1,322 eugenics papers that it had reviewed over the previous twelve months. The Eugenics Research Association boasted in 1924 that 119 of its 383 members were fellows of the American Association for the Advancement of Science, the nation's most prestigious scientific group.[18] Even the august *Encylopaedia Britannica* confidently predicted that future progress would include "the organic betterment of the race through wise application of the laws of heredity."[19]

As early as 1914, Davenport and the Eugenics Record Office had announced a platform for achieving that brighter future. One of the office's advisory groups, "The Committee to Study and to Report on the Best Practical Means of Cutting Off the Defective Germ-Plasm in the American Population," calculated that 10 percent of the American population was defective and should be sterilized.[20] It was an agenda that pleased former president Theodore Roosevelt. "At present," he wrote the committee, "there is no check to the fecundity of those who are subnormal."[21] During a national eugenics conference that year funded by John Harvey Kellogg, inventor of the flaked cereal, the scope of the needed enterprise was further defined: Over the next forty years, the country needed to sterilize 5.76 million Americans in order to reduce the percentage of defectives in the population to an acceptable level.[22]

Mendelian Madness

The scientific justification for the compulsory sterilization of the severely mentally ill rested on two premises: that "insanity" was an inherited disease and that the severely mentally ill were proficient at the mating game and thus were passing on their tainted genes to a large number of offspring. If either of these facts weren't true, then the eugenicists' argument that the mentally ill were a threat to the country's "germ plasm" would be seriously undermined.

Proving that insanity was an inherited disease fell to Aaron Rosanoff, a doctor at Kings Park State Hospital in New York. Working under Davenport's tutelage, he charted the family histories of seventy-two insane patients. His initial results were not what he expected. Among the 1,097 relatives of the seventy-two patients, only forty-three had ever been hospitalized for a mental illness—a number far too low to show a causal genetic link. Rosanoff calculated that according to Mendelian laws, 359 of the relatives should have been mentally ill. His study seemed to disprove the notion he'd set out to prove. Where had he gone wrong? The answer, he concluded, was that he had defined mental illness too narrowly. Plenty of mentally ill people were never hospitalized. "Neuropathy," he explained, manifested itself in many ways. Relatives of patients with manic-depressive insanity should be considered mentally ill if they were "high-strung, excitable, dictatorial, abnormally selfish," or if they had an "awful temper, drank periodically, [or] had severe blue spells." In a similar vein, relatives of patients hospitalized for schizophrenia should be classified as neuropathic if they were "cranky, stubborn, nervous, queer, [or] restless," if they were "suspicious of friends and relatives," if they "worried over nothing," or acted like "religious cranks." And with that neatly expanded definition of mental illness at work, Rosanoff determined that the seventy-two hospitalized patients had 351 neuropathic relatives—almost an exact match to the number needed to support his hypothesis. "The hereditary transmission of the neuropathic constitution as a recessive trait, in accordance with the Mendelian theory, may be regarded as definitely established," he happily concluded.[23]

There was—if Rosanoff's study was to be believed—a clear line separating "neuropathics" from "normals." However, Rosanoff's findings had unsettling ramifications for normals as well. Because the "neuropathy" gene was recessive, a normal person might still be a carrier of insanity, capable of passing it on. Rosanoff calculated that 30 percent of the American population was so tainted. Meanwhile, a mating between two mentally ill people, both of whom lacked the "normalcy" gene, was obviously hopeless: "Both parents being neuropathic, all children will be neuropathic."

Twenty-five years later, Boston psychiatrist Abraham Myerson pointed out how laughably bad this science was. "Whole diversities of things are artificially united. Thus, if a father has a sick headache and his descendant has dementia praecox, the two conditions are linked together in a hereditary chain."[24] Yet in the wake of Rosanoff's 1911 study, mental illness as a Mendelian disorder became the scientific paradigm presented to the public. The *Science of Eugenics,* a popular book published in 1917, told readers that "when both parents are normal but belong to insane stock, about one-fourth of their children will become insane."[25] The 1920 *Manual on Psychiatry,* a medical text edited by Rosanoff, declared, "Most of the inherited mental disorders are, like the trait of blue eyes, transmitted in the manner of Mendelian recessives."[26] Biologist Paul Popenoe, editor of the *Journal of Heredity,* explained that when an "insane" person "mates with a normal individual, in whose family no taint is found, the offspring (generally speaking) will all be mentally sound, even though one parent is affected. On the other hand, if two people from tainted stocks marry, although neither one may be personally defective, part of their offspring will be affected."[27] With such scientific dogma in mind, the *New York Times* editorialized in 1923 that "it is certain that the marriage of two mental defectives ought to be prohibited."[28]

But if proving that insanity was inherited was difficult, eugenicists had an even harder time supporting the notion that the mentally ill were prolific breeders. Even on the face of it, this seemed a dubious proposition. Schizophrenics, almost by definition, are socially withdrawn, which is just what researchers found time and again. A 1921 study determined that nearly two-thirds of males diagnosed as schizophrenic had never even had sex with a woman. Other studies found that the "insane" were less likely to be married than the general population and had mortality rates five to fifteen times those of the normal population. Even Popenoe reluctantly concluded that the insane didn't marry in great numbers and that they had so few children they didn't reproduce their own numbers. They were worse breeders, in fact, than Harvard graduates and Mayflower descendants.[29]

However, such findings didn't temper eugenicists' call for sterilizing the mentally ill. Eugenicists simply lumped them together with a larger group of misfits—the poor, criminals, and mentally handicapped—said to be siring offspring at great rates. Popenoe argued that while the mentally ill in asylums—whose lives had been the subject of the research studies—may not have been good at breeding, those in the community were making up for their shortcomings. "Mentally diseased persons who do not get into state institutions and who have not been legally labeled insane seem to have families quite as large as the average, if not larger," he said. "They are spreading defective germ plasm continually through the sound part of the community, and many of them can be pointed out with probable accuracy through a study of their ancestry."[30]

The Selling of Eugenics

During World War I, America's interest in eugenics briefly cooled as the country turned its attention to the more pressing matters of war. But the carnage of that conflict, in which the United States and European countries sent their young men to fight and die, heightened the belief, here and abroad, that societies were racially degenerating. If a society's most fit young men died in battle while the weak left at home survived to procreate, what would that do to a society's makeup in the future? With that question hanging in the air, the need for countries to adopt eugenic policies suddenly seemed more pressing.

The selling of eugenics in America began in earnest in 1921, when the American Museum of Natural History hosted the Second International Congress on Eugenics, a meeting financed in large part by the Carnegie Institution and the Rockefeller Foundation. Museum president Henry Fairfield Osborn—a nephew of J. P. Morgan—opened the session by declaring that it was time for science to "enlighten government in the prevention of the spread and multiplication of worthless members of society." Over the next few days, speakers from Johns Hopkins, Princeton, Harvard, Columbia, Cornell, MIT, and NYU, as well as other top universities, tried to do just that. They presented papers on the financial costs societies incurred by caring for defectives, the inheritability of insanity and other disorders, and the low birth rates of the elite in America. They gave talks on "The Jewish Problem," the dangers of "Negro-White Intermixture," and the "Pedigrees of Pauper Stocks." After the conference, many of the scientists' charts and exhibits were put on display in the U.S. Capitol, where they remained for three months.[31]

The meeting stirred the *New York Times* to editorialize that life, indeed, was becoming ever more unfair for the well-to-do.

> Civilization, as now organized, does not leave Nature as fresh as she has been in the past to procure the survival of the fit. Modern philanthropy, working hand in hand with modern medical science, is preserving many strains which in all preceding ages would have been inexorably eliminated. . . . While life has become easier in the lower ranges, it has become more difficult for the well born and the educated, who pay for modern philanthropy in an ever lessening ability to afford children of their own. There is a very serious question whether the twentieth century will be able to maintain and pass onward the infinitely intricate and specialized structure of civilization created by the nineteenth century.[32]

At the close of the international meeting, Davenport, Osborn, and other prominent eugenicists formed a committee to establish a national eugenics society. As a first step, they recruited a ninety-nine-member scientific advisory council, reaching out to candidates with a letter that warned of "racial deterioration" and the need for societal leaders to resist the "complete destruction" of the "white race." In a eugenic society, the letter said, "our burden of taxes can be reduced by decreasing the number of degenerates, delinquents, and defectives supported in public institutions."[33]

The advisory council, in place by 1923, was an elite group, and it remained so for the next decade. From 1923 to 1935, more than half of its members were graduates of Ivy League universities, with nearly 40 percent educated at Harvard, Yale, or Columbia. Harvard's president emeritus Charles Eliot and eight other college presidents served on the council. Professional biologists, zoologists, and geneticists made up one-third of the group. About 10 percent were psychologists. Five presidents of the American Psychological Association (past or present) were members, as were a similar number of presidents of the American Association for the Advancement of Science. Adolf Meyer, who was the leading figure in American psychiatry at that time, joined the council. So did Charles Burr, a past president of the American Neurological Association. Floyd Haviland, president of the American Psychiatric Association (APA), offered his advice as a council member. The council, which was expected to review all of the society's educational materials, represented many of the best and brightest in America—its top doctors and scientists, educated at its best universities.[34]

The American Eugenics Society (AES) was incorporated in 1926. John D. Rockefeller Jr. contributed $10,000 to help launch it. George Eastman, of Eastman Kodak fame, gave $20,000. Yale professor Irving Fisher, the best-known economist of his time, served as the first president. In a short period, it grew into a truly national organization, with chapters in twenty-eight states.

The society focused on promoting eugenics to the American public—getting textbooks and pamphlets into schools and conducting informational campaigns to build support for sterilization laws. One of its popular traveling exhibits, consisting of a board with blinking lights, was titled "Some People Are Born to Be a Burden on the Rest." Every fifteen seconds a light flashed to warn onlookers that American taxpayers had just spent another $100 caring for defectives. Every thirty seconds, a light flashed to signal that another defective had been born. At intervals of fifty seconds, a flashing light told of another criminal being carted off to prison, with the audience informed that "very few normal persons ever go to jail." Finally, after seven and one-half long minutes, a light blinked to announce that a "high grade person," at long last, had been born.[35]

State fairs proved to be particularly good forums for educating the public. In addition to its flashing-light exhibit, the society set up charts explaining Mendelian laws of inheritance and how they determined human types. "Unfit human traits," the AES advised the American public, "run in families and are inherited in exactly the same way as color in guinea pigs."[36] To further its point, the AES organized "Fitter Families" contests, with entrants submitting family histories, undergoing psychiatric exams, and taking IQ tests, all in the hope that they would be deemed Grade-A humans. Winning families joined other best-of-show livestock—pigs, goats, cows—in end-of-fair parades, the humans riding in automobiles decorated with banners proclaiming them the state's "best crop."[37]

To get the country's clergy involved, the AES sponsored annual contests with cash awards, up to $500, for ministers and priests who delivered the best eugenics sermon. In 1928, Reverend William Matson of the Methodist Episcopal Church won the top prize of $500 by telling his congregation that "modern science" had proven "all men are created unequal." With such a disparity in genetic makeup, trying to lift up the unfit with education and social programs was "like attempting to grow better alfalfa with dandelion seed." Said Matson: "We may raise a pig in the parlor but he remains a pig." Other ministers won cash prizes for telling their members that God was waiting for the human race to become "purified silver," cleansed of its "impurities of dross and alloy" and that "if marriage is entered into by those notoriously unfit to give a righteous biologic entail, the state has a right to insist on sterilization."[38]

Meanwhile, in a 137-page booklet called "Tomorrow's Children," designed to serve as the society's "catechism," schoolchildren and other readers were encouraged to think of the AES as a "Society for the Control of Social Cancer." The mentally ill and other defectives were an "insidious disease," and each time they had children, they created "new cancers in the body politic." In a modern society, cancer needed to be treated with a "surgeon's knife." At the moment, though, American society was failing to respond to this threat: "Crime and dependency keep on increasing because new defectives are born, just as new cancer cells remorselessly penetrate into sound tissue."[39]

In the 1930s, the invective from eugenicists became, in many instances, even shriller. Franz Kallman, chief of research at the New York State Psychiatric Institute, said that all people, even lovers of "individual liberty," had to agree "mankind would be much happier" if societies could get rid of their schizophrenics, who were not "biologically satisfactory individuals."[40] Charles Stockard, president of the Rockefeller Institute for Medical Research, worried that the human species faced "ultimate extermination" unless propagation of "low grade and defective stocks" could be "absolutely prevented."[41] Meanwhile, Earnest Hooton—Harvard professor of anthropology and AES council member—in his 1937 book *Apes, Men, and Morons,* compared the insane to "malignant biological growths" whose germ plasm should be considered "poisonous slime." America, he argued, "must stop trying to cure malignant biological growths with patent sociological nostrums. The emergency demands a surgical operation."[42]

All of this was a far cry from the sentiments that had governed moral treatment a century earlier. In the first decades of the twentieth century, the American public regularly heard the insane likened to "viruses," "social wastage," and "melancholy waste products." They were a plague on civilization, one that in nature would have been quickly eliminated. Scorn toward the severely mentally ill had become the popular attitude of the day, and that attitude was the foundation for laws that curbed their right to pursue, as the Declaration of Independence had once promised, life, liberty, and happiness.

First Detention, Then Sterilization

From the beginning, American eugenicists had a clear-cut agenda for preventing the mentally ill from having children. States would need to make it illegal for the insane to marry, segregate them into asylums, and release them only after they had been sterilized. Only then would they cease to be a threat to the country's genetic makeup.

The "insane" began to lose the right to marry in 1896, when Connecticut became the first state to enact such a prohibition. North Dakota quickly followed suit, as did Michigan, which threatened the insane with a $1,000 fine and five years in prison should they dare to wed. By 1914, more than twenty states had laws prohibiting the insane from marrying, and, in 1933, Popenoe matter-of-factly reported that there were no states left where the insane could legally tie the knot. Yet few eugenicists believed that such laws did much good. Not only did they fail to stop people from having children out of wedlock, they weren't even very effective at stopping marriage. Insane people, Popenoe said, when applying for marriage licenses, were tempted to lie and claim that they were quite well. "The candidate," he explained, "might be prejudiced in his own favor."[43]

Segregating the insane in asylums promised to be much more effective. In fact, this was the first goal of eugenicists, ahead even of sterilization. In its 1914 report on cutting off "defective germ plasm" in the American population, the Eugenics Record Office noted that sterilization "is simply an insurance when segregation ceases."[44] That same year, at a national eugenics conference, Wisconsin professor Leon Cole argued that "it is coming, I think, to be generally conceded that permanent segregation, at least during the period of reproductive capacity, is going to prove the most feasible if not the most effective of restrictive measures."[45] There was no talk, among the eugenicists, of sending the mentally ill to hospitals for therapeutic purposes. Instead, they envisioned sending the mentally "unfit," in essence, to detention camps, run on bare-bones budgets, with the "patients" kept there until they had passed reproductive age or had been sterilized.

To a surprising degree, eugenicists were successful in achieving this goal. In 1880, before the eugenics spirit began to take hold, America's asylums held 31,973 people, or 0.06 percent of the population. By 1929, 272,527 people were in mental hospitals—or 0.23 percent.[46] The ratio had increased fourfold in fifty years. In 1923, a *Journal of Hereditary* editorial concluded, with an air of satisfaction, that "segregation of the insane is fairly complete."[47]

The third aspect of the eugenicists' agenda, compulsory sterilization of "defectives," took longer for Americans to endorse. As early as 1882, a year before Galton coined the term "eugenics," William Goodell, a well-known gynecologist in Pennsylvania, had proposed castrating the mentally ill in order to prevent them from bearing "insane offspring." Goodell, who reported that surgical removal of a woman's ovaries could cure "ovarian insanity," predicted that in a "progressive future," it would "be deemed a measure of sound policy and of commendable statesmanship to stamp out insanity by castrating all the insane men and spaying all the insane women."[48] His views were echoed by F. D. Daniel, editor of the *Texas Medical Journal*, who believed that castrating insane men would also keep them from masturbating and thus "would be an advisable hygienic measure."[49]

Despite such sentiments from physicians, castration was a surgery that evoked shudders in the general population, too extreme to be written into law. In the 1890s, however, medical procedures for sterilizing men and women without castration were developed, which put this possibility into a new light. All that had to be done was to cut the vas deferens in men or tie a woman's fallopian tubes, neither of which prevented people from having sex. With such a minor surgery available, who could protest against its use on the insane? "It is the acme of stupidity to talk in such cases of individual liberty, of the rights of the individual," said New Jersey urologist William J. Robinson, a well-known eugenics advocate. "Such individuals have no rights. They have no right in the first instance to be born, but having been born, they have no right to propagate their kind."[50]

In 1907, Indiana became the first state to pass a compulsory sterilization law. It did so in the name of science, the bill stating that heredity had been shown to play a dominant role in the "transmission of crime, idiocy, and imbecility." Over the next two decades, thirty state legislatures approved sterilization bills, and repeatedly they did so based on an argument that science had proven that defectives breed defectives. Their lists of degenerate hereditary types were often long. In its 1913 bill, Iowa said that those in need of sterilization included "criminals, rapists, idiots, feeble-minded, imbeciles, lunatics, drunkards, drug fiends, epileptics, syphilitics, moral and sexual perverts, and diseased and degenerate persons"—a catch-all description, in essence, for people viewed as social "scum" by the legislature.

Despite the enthusiasm of state legislatures for such measures, states—with the notable exception of California—did not begin sterilizing their "defectives" in any great numbers, at least not until 1927. Opponents, which naturally included Catholics and non-English immigrant groups, argued that such laws violated constitutional safeguards against cruel and unusual punishment, due process of law, and equal protection of laws—the last flaw arising because the bills regularly authorized sterilization only of institutionalized people, as opposed to all people with supposed hereditary defects. By 1923, laws in Iowa, New York, New Jersey, Nevada, Michigan, Indiana, and Oregon had been declared unconstitutional in state courts. Governors in Pennsylvania, Oregon, Vermont, Idaho, and Nebraska vetoed sterilization bills, with the most stinging and wittiest rebuke coming from Pennsylvania's Samuel Pennypacker. Rising to speak at a dinner after his veto, he was roundly greeted with boos, catcalls, and sneering whistles. "Gentlemen, gentlemen," he implored, raising his arms to silence the crowd of legislators, "you forget you owe me a vote of thanks. Didn't I veto the bill for the castration of idiots?"[51]

As a nation, America was having a difficult time making up its mind about sterilization. From 1907 to 1927, about 8,000 eugenic sterilizations were performed—a significant number, yet only a tiny percentage of the people confined in asylums. Was it constitutional or not? Was this a practice consistent with the governing principles of the country?

In 1927, the U.S. Supreme Court—by an 8–1 majority in the case of *Buck v. Bell*—ruled that it was. In his written opinion, Oliver Wendell Holmes supported the decision by noting "experience has shown that heredity plays an important part in the transmission of insanity, imbecility, etc." Bad science had become the foundation for bad law:

> We have seen more than once that the public welfare may call upon the best citizens for their lives. It would be strange if it could not call upon those who already sap the strength of the state for these lesser sacrifices, often not felt to be such by those concerned, in order to prevent our being swamped with incompetence. It is better for all the world, if instead of waiting to execute degenerate offspring for crime, or to let them starve for their imbecility, society can prevent those who are manifestly unfit from continuing their kind.[52]

At that moment, America stood alone as the first eugenic country. No European nation had enacted a statute for compulsory sterilization of the mentally ill and other misfits. In the wake of the U.S. Supreme Court decision, the number of eugenic sterilizations in the United States markedly increased, averaging more than 2,200 annually during the 1930s. Editorials in the *New York Times* and leading medical journals like the *New England Journal of Medicine* spoke positively about the practice. A 1937 *Fortune* magazine poll found that 66 percent of Americans favored sterilizing "defectives." By the end of 1945, 45,127 Americans had been sterilized under such laws, 21,311 of whom were patients in state mental hospitals.

A Humanitarian Therapy

Although the U.S. Supreme Court spoke of sterilization as a small "sacrifice" that the unfit should make for the national good, no society likes to perceive itself as mean-spirited toward its misfits. Nor do physicians want to see themselves as implementers of social policy that might harm their patients. They want to provide care that is helpful. Those two needs, for society to view itself in a good light and for physicians to view themselves as healers, were revealed early on in California, where, by the end of World War II, nearly 50 percent of all sterilizations of the mentally ill in the United States had been performed. There, physicians came to view sterilization as providing patients with a *therapeutic* benefit, one that, or so society was told, evoked gratitude in most patients.

California approved its Asexualization Act in 1909, a law pushed by a physician, F. W. Hatch, who was then named superintendent of the state's mental hospitals. He promised to use the law to ensure that asylum "defectives should leave behind them no progeny to carry on the tainted and unhappy stream of heredity."[53] Two amendments to the law, in 1913 and 1917, broadened the definition of who was to be considered defective and authorized the state to sterilize such people without their consent. By 1921, nearly 80 percent of the 3,233 eugenic sterilizations done in the United States had been performed in California.

As California doctors conducted such operations, they constructed various rationales to explain why sterilization benefited the mentally ill. In the male, a number of California doctors reasoned, the operation allowed for the conservation of sperm, which should be considered the "elixir of life." "By this interruption in the continuity of the vas," explained Fred Clark, superintendent at Stockton State Hospital from 1906 to 1929, "the testicular secretion is absorbed. Since performing these operations we are led to believe, by the improvement in mental and general health, that there is a definite beneficial effect from the absorption of the testicular secretion." Other physicians speculated that the mentally ill had abnormal testicles to begin with, similar in size and appearance to the testicles of older men, and thus were in particular need of being rejuvenated through the snipping of the vas deferens. As one Stockton physician wrote: "The greatest benefit seems to occur in cases of Dementia Praecox and I believe that there has been a sufficient number of cases improved to warrant calling the operating a therapeutic measure." Women who were sterilized, meanwhile, were said to benefit psychologically from the operation. They no longer had to fear getting pregnant and going through the rigors of childbirth and motherhood.[54]

Having fashioned a therapeutic view of sterilization, asylum physicians in California could comfortably pitch it to their patients and to their patients' families, seeking their consent, even though the law did not require it. It was, after all, a procedure that would help the mentally ill get well. In one letter to a family, a Stockton physician wrote: "There is comparatively little or no danger in the operation and many of our patients have shown a *marked improvement.* Under the circumstances I think it is advisable in this case." Here is a 1928 transcript of a doctor explaining the operation to a patient:[55]

Doctor: Have you ever been sterilized?

Patient: No.

Doctor: You had better let us operate on you while you are here.

Patient: Doctor, will that bring better composure to the nervous system?

Doctor: It is supposed to, it has in a number of cases, we do not guarantee it, but in a number of cases it has had marked beneficial effects. It cannot hurt you and does not interfere with your sexual life in any way, we just cut a little duct and you absorb your own secretions.

Patient: It has always been all right with me.
Doctor: Well, it cannot hurt you and it might have a marked beneficial result.
Patient: I will be very much obliged to you, sir.

In this interplay between doctor, family, and patient, a story of humanitarian care was being woven. Doctors found sterilization to be therapeutic; the mentally ill desired it. In 1929, the Human Betterment Foundation—a state eugenics organization led by wealthy banker Ezra Gosney and Popenoe—reported that 85 percent of the sterilized mentally ill were either thankful for the operation or at least indifferent to it. Many women, they said, were "pathetic in their expression of gratitude and their wish that other women who faced the combination of pregnancy and psychosis might have the same protection."[56] The California Department of Mental Hygiene even began to list sterilization as a medical treatment that was provided to patients in its state hospitals. This good news tale convinced the public: In 1935, 83 percent of all Californians favored eugenic sterilization of the mentally ill.[57]

The voice of the mentally ill so sterilized is almost entirely absent from the medical literature. There is, however, one faint lament that can be heard today. Eight years after being sterilized, a twenty-nine-year-old man described to Popenoe how his life had been forever changed:

> I was operated on in 1918 when I was 21. I was a patient for some 3 1/2 months. Will say this, that it was all a mistake . . . I would rather not be sterilized as I do not think there is the slightest danger of myself being responsible for any weak or feebleminded children, and I shall ever bemoan the fact that I shall never have a son to bear my name, to take my place, and to be a prop in my old age.[58]

That, in the era of American eugenics, was the cry of the "insane."

The Killing Fields

America's embrace of eugenic sterilization as a progressive health measure had consequences for the mentally ill in other countries as well. Two years after the U.S. Supreme Court deemed it constitutional, Denmark passed a sterilization law, and over the next few years, Norway, Sweden, Finland, and Iceland did too. America's influence on Nazi Germany was particularly pronounced, and it was in that country, of course, that eugenics ran its full course.

Prior to World War I, eugenics was not nearly as popular in Germany as it was in the United States. Germany's parliament defeated a sterilization bill in 1914, and the country didn't pass any law prohibiting the mentally ill from marrying. However, after the war, eugenics gained a new appeal for the German population. Germany's economy lay in ruins after the war, and more than 1.75 million of its ablest young men had died in the conflict. How could the impoverished country afford the cost of caring for "defectives" in asylums? Should the unfit be allowed to pass on their tainted genes while so many of its healthy young men had died before having a chance to become fathers? In 1925, Adolf Hitler, in *Mein Kampf*, hailed eugenics as the science that would rebuild the nation. The state, he wrote, must "avail itself of modern medical discoveries" and sterilize those people who are "unfit for procreation."

Much as U.S. geneticists had, German eugenicists sought to develop scientific evidence that mental illnesses were inherited and that such genetic disease was spreading through its population. American money helped fund this effort. In 1925, the Rockefeller Foundation gave $2.5 million to the Psychiatric Institute in Munich, which quickly became Germany's leading center for eugenics research. In addition, it gave money to the Kaiser Wilhelm Institute for Anthropology, Human Genetics and Eugenics in Berlin, which was used to pay for a national survey of "degenerative traits" in the German population.[59]

After Hitler came to power in 1933, Germany passed a comprehensive sterilization bill. The German eugenicists who drew up that legislation had gone to school on the U.S. experience, which American eugenicists noted with some pride. "The leaders in the German sterilization movement state repeatedly that their legislation was formulated only after careful study of the California experiments," wrote Margaret Smyth, superintendent of Stockton State Hospital, after touring Germany in 1935. "It would have been impossible they say, to undertake such a venture involving

1 million people without drawing heavily upon previous experience elsewhere."[60]

Many in Germany and in the United States also saw the Nazi bill as morally superior to any U.S. state law, as it had elaborate safeguards to ensure due process. German physicians were required to report "unfit" persons to Hereditary Health Courts, which then reviewed and approved patients for sterilization. There were even provisions for appeal. This was an example of how one country could learn from another and push modern medicine forward. Germany, the *New England Journal of Medicine* editorialized, had become "perhaps the most progressive nation in restricting fecundity among the unfit." The American Public Health Association praised Germany in similar terms and at its annual meeting in 1934 mounted an exhibit on Germany's sterilization program as an example of a modern health program. The *New York Times*, meanwhile, specifically sought to "dispel fears" that Hitler, with his new sterilization law, was pursuing "a discredited racial idea." Germany, it wrote, was simply following in the path of other "civilized" countries, most notably the United States, where "some 15,000 unfortunates have been harmlessly and humanely operated upon to prevent them from propagating their own kind."[61]

Over the next six years, Germany sterilized 375,000 of its citizens. The pace of eugenic sterilization during this period picked up in the United States as well, and the Scandinavian countries also sterilized a number of their "defectives." A eugenic treatment born in the United States had spread into a half dozen European countries. However, Germany was employing it with a fervor missing in the United States, which led some American eugenicists to fret that Hitler was now "beating us at our own game." While America was "pussy-footing around" with the procedure, complained Leon Whitney, field secretary for the American Eugenics Society, Germany was making "herself a stronger nation."[62]

And then Nazi Germany took eugenic treatment of the mentally ill to its ultimate end.

Eugenic attitudes toward the mentally ill—that they were a drain on society and a threat to its "germ plasm"—inevitably raised the possibility of a more extreme measure. Should a state simply kill its insane? This question was first raised in the United States in

1911, when Charles Davenport published *Heredity in Relation to Eugenics.* Although he generally argued against killing the unfit, he wrote that if a society had to choose between allowing "mental defectives" to procreate and killing them, the latter would be the preferable alternative. "Though capital punishment is a crude method of grappling with the difficulty [of defectives]," he concluded, "it is infinitely superior to that of training the feeble-minded and criminalistic and then letting them loose upon society and permitting them to perpetuate in their offspring these animal traits."[63] Five years later, Madison Grant, a wealthy New York lawyer and a founder of the American Eugenics Society, pushed this notion a step further in his book *The Passing of the Great Race.* "The Laws of Nature require the obliteration of the unfit, and human life is valuable only when it is of use to the community or race," he argued. "A great injury is done to the community by the perpetuation of worthless types."[64]

The idea that the mentally ill, and other misfits, were "useless eaters" was now alive and loose in the Western world. Grant's best-selling book went through four editions and was translated into French, Norwegian, and German. Hitler, according to German historian Stefan Kühl, later wrote Grant a fan letter, telling him "the book was his Bible."[65]

Over the next two decades, the notion that state killing of the mentally ill might be acceptable popped up in various forums in the United States. In 1921, Connecticut legislators, having toured the State Hospital for the Insane in Norwich, where they observed a fifty-year-old man manacled to an iron bed, contemplated passing a law "that would provide that persons found to be hopelessly insane after observation and examination of experts should be put to death as mercifully as possible, preferably by poison." The *New York Times* headline proclaimed that the man had been "Exhibited as Case for Merciful Extinction."[66] The hateful rhetoric of American eugenicists in the 1920s and 1930s, which characterized the mentally ill as "social wastage," "malignant biological growths," and "poisonous slime," also implicitly suggested that perhaps society should find a way to get rid of them. The insane, explained Harvard's Earnest Hooton, were "specimens of humanity who really ought to be exterminated."[67] Finally, in 1935, Alexis Carrel, a Nobel Prize–winning physician at Rockefeller Institute for Medical Research in New York City, made the point explicit. In his book *Man the Unknown*, he wrote:

> Gigantic sums are now required to maintain prisons and insane asylums and protect the public against gangsters and lunatics. Why do we preserve these useless and harmful beings? The abnormal prevent the development of the normal. This fact must be squarely faced. Why should society not dispose of the criminals and insane in a more economical manner? . . . The community must be protected against troublesome and dangerous elements. How can this be done?

Carrel answered his own question. The insane, or at least those who committed any sort of crime, "should be humanely and economically disposed of in small euthanasic institutions supplied with proper gases."[68]

Nazi Germany began killing its mentally ill with "proper gases" in January 1940. It did so based on a simple eugenics rationale: Four months earlier, it had invaded Poland, and killing the mentally ill promised to free up hospital beds for the wounded, and also spare the state the expense of feeding them. Over the course of eighteen months, the Nazis gassed more than 70,000 mental patients. Program administrators even calculated the resultant financial benefits, carefully itemizing the food—bread, margarine, sugar, sausage, and so on—no longer being consumed by those who had been killed. Hitler called a halt to this systematic killing of the mentally ill on August 24, 1941; the gas chambers were dismantled and sent to concentration camps in the East, where they were reassembled for the killing of Jews and others "devoid of value." A path that had begun seventy-five years earlier with Galton's study of the superior traits of the ruling English elite, and had then wound its way through the corridors of American science and society, had finally arrived at Auschwitz.

America's Concentration Camps

Although Americans had learned of Nazi concentration camps early in World War II by reading about them in newspapers and magazines, the full horror of those prisons did not hit home until photographs of the camps appeared. When Allied troops liberated the camps in 1945, America and the rest of the world were confronted with the images that so seared the twentieth-century mind. Jews of all ages in striped prison garb, emaciated, their eyes bewildered—it all spoke of unfathomable suffering.

Shortly after the war ended, Americans found themselves staring at photographs of a lost world closer to home. First in *Life* magazine and then in a book by journalist Albert Deutsch, America was given a vivid tour inside its state mental hospitals. The pictures seemed impossible: Mentally ill men huddled naked in barren rooms, wallowing in their own feces; barefoot women clad in coarse tunics strapped to wooden benches; sleeping wards so crowded with threadbare cots that patients had to climb over the foot of their beds to get out. One photo caption told of restrained patients, unable to use their hands, lapping food from tin plates, like dogs eating from bowls. In *The Shame of the States*, Deutsch drew the inevitable comparison:

> As I passed through some of Byberry's wards, I was reminded of the Nazi concentration camps at Belsen and Buchenwald. I entered buildings swarming with naked humans herded like cattle and treated with less concern, pervaded by a fetid odor so heavy, so nauseating, that the stench seemed to have almost a physical existence of its own. I saw hundreds of patients living under leaking roofs, surrounded by moldy, decaying walls, and sprawling on rotting floors for want of seats or benches.[69]

Numerous newspapers ran scathing exposés as well. Papers in Norman, Oklahoma; Cleveland; Miami; Baltimore—their reports all told a similar story. In hospital after hospital, scenes of patients cuffed, strapped to chairs, and wrapped in wet sheets. Facilities infested with rats, cockroaches, and other vermin. Patients, the reporters noted, went weeks, months, and even years without seeing a doctor. Order in the madhouses was maintained by attendants who, with some frequency, beat unruly patients. The mentally ill in such hospitals, concluded *Life* writer Albert Maisel, were "guiltless patient-prisoners."[70]

At the time, *Life* and other publications blamed the shameful conditions on public neglect and penny-pinching legislators. It was shameful, but not a willful act. In a sense, that was true. The Great Depression in the 1930s and the stresses of World War II had taken their toll. However, the deterioration of the state mental hospitals was also consistent with eugenic beliefs. The mentally ill needed to be segregated, and "normal" society, burdened with this expense, needed to keep this cost to a minimum. The same skimping on funds for the mentally ill occurred in Germany after Hitler assumed power in 1933. And if the magazines and newspapers had looked back at the decline of state asylums from 1900 to 1945, they would have seen that it occurred in lockstep with the rise of the eugenics movement.

At the turn of the century, there were 126,137 patients in 131 state asylums. Forty years later, there were 419,374 patients in 181 state hospitals. The average patient census had grown from 962 to 2,316; a few hospitals housed more than 4,500 people. However, the asylums were not filling up with an increased number of "insane" patients. Society was dumping all kinds of "misfits" into the institutions—alcoholics, epileptics, vagrants, the senile elderly, drug addicts, syphilitics, and the mentally ill. They were lockups for the "social wastage" said by eugenicists to be plaguing modern societies; by some estimates, fewer than 50 percent of the people committed to asylums in the 1930s were ill with schizophrenia, manic depression, or other well-defined forms of "insanity."[71]

During this period, funding for the state asylums, on a per-patient basis, became ever more parsimonious. By the early 1940s, states were spending less than $1 per day for each asylum patient, which, on an inflation-adjusted basis, was less than one-third what Pennsylvania Hospital had spent on its patients in 1860. It was also only one-eighth the amount spent by private sanitariums in the 1940s. The American Psychiatric Association estimated at that time that it took at least $5 per day to provide patients with decent care. And with state hospitals operating on such bare-bones budgets, death rates for asylum patients soared. In the 1930s, patients in New York state hospitals died at five times the rate of the general population, and mortality rates were particularly high for young people twenty to twenty-four years old—the very group that eugenicists did not want to see breed. Five percent of this young mentally ill group died annually, a mortality rate fifteen times higher than the rate for people of the same age in the general population.[72] Deutsch later described the poor care as "euthanasia" through "neglect."[73]

Finally, just as the eugenicists had urged, the asylums were increasingly run as places of confinement—facilities that served to *segregate* the misfits from society—rather than as hospitals that provided medical care. At the turn of the century, state asylums reported having one doctor for every 100 to 175 patients. By 1943, state hospitals averaged one doctor for every 277 patients, and only one nurse for every 176 patients.[74] With so few doctors and nurses present, the patients' daily lives were largely controlled by poorly paid attendants, who often had to live on site. *Life,* in its 1946 exposé, reported that attendants in the Pennsylvania state hospitals earned $900 a year, less than half the $1,950 paid to state prison guards, even though, the magazine wrote, "the psychiatric attendant's job is more dangerous and certainly far less pleasant that that of the prison guard." In the pecking order of social discards, asylum patients fell below criminals.

As in prisons, the attendants' main job was to maintain order, which they did with the liberal use of sedatives and restraints, and, if necessary, with force. As hitting patients was usually against the rules—they were, after all, theoretically working in hospitals—they developed methods for beating patients in ways that didn't leave visible bruises. Marle Woodson, a newspaper reporter who was hospitalized in an Oklahoma asylum in 1931, told of two common methods:

> Wet toweling a patient is choking him unconscious by getting a wet towel around his neck and twisting on it from the back until he succumbs. Sometimes a pillowslip is used instead of a towel . . . Soaping a man down means knocking him down with a slug made of a hard bar of soap in the toe of a sock. Such a slug will knock a patient down, often rendering him unconscious, without leaving telltale marks. The wet towel treatment also leaves no marks or scars for inquisitive officials, hospital authorities or unexpected visitors to find. Hurrah for the inventive genius of younger America. It finds ways to make itself safe.[75]

During the 1930s, the deteriorating conditions in the mental hospitals were often discussed, a concern of many state and federal agencies. But the public discussion usually centered on how the hospitals could be improved, and not on whether the country was, in fact, not running hospitals at all, but simply locking up its mentally ill in penal camps. In 1933, the American Medical Association (AMA) was provided with evidence of this alternative reality, but rather than publicly confront this truth, it chose instead to cover it up.

Two years earlier, the association had hired a young physician, John Grimes, to investigate the country's mental hospitals. He sent surveys to 174 state hospitals, and either he or one of his staff personally visited nearly all of them. He came back with an unexpectedly disturbing portrait. On the outside—the facade that was being presented to the public—the state mental hospitals looked to be in good shape. Their grounds, Grimes said, had a beauty that "approaches that of city parks, with shade, grass, flowers, streams, rustic bridges, pavilions, walks, baseball diamonds, miniature golf links, and tennis and croquet courts." Inside, however, was a different story. Hospitals were so crowded that patients were sleeping in hallways, dining rooms, living rooms, gymnasiums—any place that a cot could be set up. He even found instances of disturbed patients having to sleep two to a bed. Attendants, he said, acted like prison guards; wards were locked and the windows barred. As for feeding the patients, the hospitals often had to rely on what could be reaped from their own farms. Eggs were a rare delicacy; few could give patients milk to drink. The primary purpose of such institutions, Grimes concluded, was not medical but "legal." They served to confine people unwanted by society, including many who were not mentally ill but were there "because of unsocial or antisocial manifestations."[76]

In essence, Grimes had discovered that the nation was deluding itself. But it wasn't a message that the AMA wanted to hear. The AMA told him to change his report; he refused, and was fired. At its annual meeting, the AMA circulated a brief ten-page summary of the survey statistics Grimes had gathered but pointedly omitted all of the damning eyewitness accounts in Grimes's report, and then it let the matter drop. Grimes had to self-publish his findings, and without the backing of the AMA, his book attracted little public notice. The AMA, he angrily wrote, had "ignored an opportunity to plead the case of America's most neglected and most helpless group of hospital patients."

Patients were left to plead their own case. And in their writings—a handful published their stories during the 1930s and 1940s—they spoke most bitterly of the hypocrisy of it all. They were locked up in pitiful asylums, and yet they would read in magazines and newspapers about how psychiatrists, at their annual meetings, boasted of therapeutic advances, or about how government agencies were working to provide better care in the mental hospitals—it was all the stuff of a societal and medical fantasy. There was, wrote Harold Maine, in his book *If a Man Be Mad,* nothing one could do to "prod the nation into an awareness of the way it had been duped with the folklore about modern institutional psychiatry."[77]

The curtain was finally raised—more than a decade after the AMA covered up its own report—by an unusual group of attendants: nearly 3,000 conscientious objectors to the war, who had chosen to work in mental hospitals as an alternative form of service. They took their eyewitness stories to district attorneys, to local reporters, and to Albert Maisel at *Life,* and they also published their own damning book, *Out of Sight Out of Mind.* In 1944, an Ohio grand jury investigating conditions at Cleveland State Hospital, where several patients had died after being beaten with belts, key rings, and metal-plated shoes, summed up the state of affairs: "The atmosphere reeks with the false notion that the mentally ill are criminals and subhumans who should be denied all human rights and scientific medical care."[78]

And then, the Ohio panel issued a stunning indictment:

> The grand jury is shocked beyond words that a so-called civilized society would allow fellow human beings to be mistreated as they are at the Cleveland State Hospital. . . . We indict the uncivilized social system which in the first instance has enabled such an intolerable and barbaric practice to fasten itself upon the people and which in

> the second instant permits it to continue . . . The Grand Jury condemns the whole socio-political system that today allows this unholy thing to exist in our State of Ohio.

At least in that courtroom, eugenic attitudes toward the mentally ill in the United States had, at long last, been heartily denounced.

4

TOO MUCH INTELLIGENCE

I think it may be true that these people have for the time being at any rate more intelligence than they can handle and that the reduction of intelligence is an important factor in the curative process. I say this without cynicism. The fact is that some of the very best cures that one gets are in those individuals whom one reduces almost to amentia [simple-mindedness].

—Dr. Abraham Myerson[1]

ALTHOUGH LEADING AMERICAN psychiatrists may have supported eugenic policies, the eugenics agenda as a whole was driven primarily by people outside medicine. Davenport, Grant, Popenoe—none were doctors. As a group, American psychiatry was rather ambivalent about the whole affair, at times embracing state sterilization laws and at other times quietly questioning the science. Yet eugenics provided a societal context for asylum medicine, and that context dramatically influenced the type of medical therapeutics that were adopted in the 1930s for psychotic disorders. At that time, psychiatry embraced a quartet of therapies—insulin coma, metrazol convulsive therapy, electroshock, and prefrontal lobotomy—

that all worked by damaging the brain. And from there, one can follow a path forward to the therapeutic failure documented by the World Health Organization in the 1990s, when it determined that schizophrenia outcomes were much better in the poor countries of the world than in the United States and other "developed" nations.

Prior to the introduction of the four treatments just mentioned, asylum psychiatry spent decades experimenting with physical remedies of every type. With the demise of moral therapy in the late 1800s, psychiatry had vowed to turn itself into a scientific discipline, and for all intents and purposes, that meant finding physical, or somatic, treatments for psychotic disorders. Although Freudian theories of the mind grabbed the imagination of American psychiatrists in the early 1900s, psychoanalysis was never seen as particularly useful or practical for treating institutionalized patients. The Freudian couch was seen as a method for treating neurotic patients in an office setting. Asylum psychiatry kept its sights set on finding somatic therapies that could be quickly applied and that would "work" in a quick manner as well.

The reform vision articulated by leaders of American psychiatry in the 1890s was well reasoned. Medical schools, they argued, would need to teach asylum medicine as part of their curriculums. Research laboratories for conducting pathological investigations into the biological causes of insanity would have to be established. It was hoped that the knowledge to be so gained would then lead to treatments that helped correct that abnormal biology. It all made perfect sense, as this was the research paradigm that was leading to such notable progress in general medicine. In the 1880s, the organisms that caused tuberculosis, cholera, typhoid, and diphtheria had been isolated; antitoxins for typhoid and diphtheria were then developed that greatly reduced mortality rates from those two diseases. A scientific approach to illness could clearly produce great results.

However, as psychiatry sought to remake itself in this way, it was also being chased by its own internal devils. The stinging attacks by neurologists had left the public convinced that asylum doctors were incompetents, or worse. Asylum medicine, a *Nation* writer had sneered, was the "very worst" department in all of medicine.

Psychiatry had a palpable need for a therapeutic triumph, one that would rescue its public image and provide a balm for its own inferiority complex. And with that emotional need spurring it on, psychiatry was primed to shortcut the research process and skip straight ahead to the part about announcing therapeutic success. This, in fact, began to happen almost from the moment that the leaders of asylum psychiatry laid out their plans for reform, so much so that the editors of the *American Journal of Insanity* could happily report in 1896 that the "present summary (of published articles) is an almost unbroken record of medical progress." In particular, the journal noted, hydrotherapy was producing "remarkable results" that "would have been impossible to get by the old method of treatment."[2]

With such claims appearing in the medical literature, hydrotherapy quickly came to occupy a central place in asylum medicine's armamentarium. Private sanitariums and better-funded state hospitals made their hydrotherapeutic units, with their rows of bathtubs and gleaming plumbing, into clinical showpieces that they proudly presented to the public. At first glance, several asylum doctors admitted, it was difficult for the medically untrained eye to see just what was so new about the water therapies. Warm baths, touted for their soothing effects, seemingly recalled the ministrations of the York Quakers. Other versions of hydrotherapy, such as the continuous bath and needle shower, appeared less benign and looked suspiciously like the discredited therapies of old for restraining, depleting, and punishing patients. But such similarities, asylum doctors assured the public (and each other), were only skin deep.

The prolonged bath involved strapping a disruptive patient into a hammock suspended in a bathtub, with the top of the tub covered by a canvas sheet that had a hole for the patient's head. At times, cold water would be used to fill the tub and at other times, water that felt hot to the touch. Patients would be kept there for hours and even days on end, with bandages sometimes wrapped around their eyes and ears to shut out other sensations. Ice caps were occasionally applied to their heads as well. Although it appeared simply to be an updated version of Rush's tranquilizer chair, asylum doctors carefully explained in their medical journals why such an extended stay in the tub was good for the patient. The continuous bath, they said, acted as a "water jacket" that "induces physiological fatigue without the sacrifice of mental capacity" and stimulates "the excretory function of the skin and kidneys." In their reports, they even provided detailed statistics on how the prolonged baths changed body temperature, respiration, and red blood-cell counts—evidence that the continuous bath was a carefully tested remedy for mental illness. They were also meticulous about detailing the risks of this medical treatment. Heat stroke, heat exhaustion, and "occasional scaldings" had been known to occur. All in all, though, reported Edward Strecker, a prominent psychiatrist at the Pennsylvania Hospital for the Insane, in 1917, patients could be kept in continuous baths "weeks, or even months, without untoward results." He advised putting pictures on the bathroom walls, making it a more pleasing environment for the patient, as the tub room should be considered a "living apartment."[3]

The needle shower, or jet douche as it was sometimes called, consisted of pummeling the patient with pressurized water. Various "prescriptions" for such showers called for dialing up pressures to forty pounds, with water temperatures as chilly as 50° Fahrenheit. The carefully timed cold showers would last a minute or two. The pounding was said to provide a variety of physiological benefits, such as stimulating the heart, driving blood to the internal organs, and inducing "glandular action by its tonic effect on the general cutaneous circulation." It was reported to be particularly useful for rousing depressed patients. But as one physician acknowledged, "we meet with more or less opposition on the part of the patient to the administration of these baths."[4]

The water therapy most reviled by patients was the wet pack. Attendants would dip sheets into either cold or hot water, then wrap them tightly around the patient "so that he cannot move anything except his head, fingers, and toes." A woolen blanket might then be pinned to the sheets, and, at times, the entire bundle tied to a bed. Patients would be left trussed up in this manner for hours and, at times, even for a day or two, abandoned in these extended treatments to wallow in their feces and urine. But that was the least of their discomfort. As the sheets dried, they would shrink tightly about the patients. With their bodily heat so snugly retained, they would experience an awful sensation of burning up, and of suffocation. Many struggled mightily to escape, so much so that "cardiac collapse" was an admitted risk. As one patient said at a 1919 hearing on conditions in California asylums, "You are in a vice, and it is inhuman treatment."[5]

However, asylum doctors saw wet packs through a different prism. In their writings, they took great pains to distinguish them from the cuffs, mitts, camisoles, and tranquilizer chair of yore. "It must appear to many that the chief object of the pack is restraint," admitted Boston's Herman Adler, "[yet] nothing can be further from the truth . . . it is a valuable therapeutic measure." The wet pack, he explained, was a physiologically beneficial treatment for "restlessness." The excited patient tended to lose bodily heat, and this necessitated the use of the wet pack to "conserve the body temperature." Once the patient had been quieted and drained by the wet pack, the patient could be treated with the prolonged bath, which would "prevent the evaporation of water from the skin," providing further conservation of the patient's body heat. Restraint was decidedly not the aim of the wet pack, he concluded; rather it was simply a means of "applying a therapeutic agent without the cooperation or even the consent of the patient."[6]

Others echoed Adler's beliefs. "Hydrotherapy," said one nurse, testifying at the 1919 California investigation, "is the only scientific treatment for the acute excitement of the insane that has yet been discovered."[7] Indeed, this was the very somatic therapy that, in the eyes of many, separated modern hospitals from asylums of old. As Allen Jackson, chief physician at the Philadelphia Hospital for the Insane, rather huffily noted in the *Journal of the American Medical Association:* "'Lunatic asylum' is the proper nomenclature for an institution which has no hydrotherapy unit; to call such an institution a hospital would be a misnomer and, to say the least, exceedingly out of place."[8]

A Bounty of Remedies

In the first decades of the twentieth century, hydrotherapy was the one somatic treatment that was widely practiced. Beyond that, physical therapies came and went with great rapidity. Remedies of every kind and stripe were tried, as hardly any hypothesis was seen as too outlandish not to test. As physicians did so, they invariably reported good results, tallying up impressive numbers of cures, remissions, and improvements. Rarely did anyone conclude that his novel therapy provided no benefit at all. There would typically be a period of enthusiasm for the therapy that was soon followed by disappointment as others tried it and found its merits to be less compelling.

Early on, during the 1890s and the first decade of the twentieth century, gynecological surgeries—for purposes other than eugenic sterilization—enjoyed a certain vogue. Such treatment arose partly from Victorian attitudes toward sexuality, and partly from the maturation of gynecology as a medical specialty. Just as neurologists had looked at the great numbers of hospitalized mentally ill as a rich source of patients, so did gynecologists. Many were so avid in their enthusiasm for curing insanity by surgically removing the uterus or ovaries that the American Medico-Psychological Association, in the early 1890s, had to caution against overuse of this remedy. Even so, for the next fifteen years, various gynecologists continued to claim that hysterectomies and ovariectomies produced improvement in more than 50 percent of their insane female patients. "The gynecologist," proclaimed W. O. Henry, at the 1906 annual meeting of the American Medical Association, "may cure various forms of insanity if [pelvic] irritation is entirely removed . . . by whatever means are necessary, no matter how radical the [surgical] work required."[9]

Much attention also focused on the pathological influence that the vagina and the nerve-rich clitoris could have on the female mind. Women, said one physician, "are deeply concerned about these organs," and "insanity may occur because their minds are very much agitated" by this undue concern.[10] Direct evidence of a female mind led astray could sometimes be found through measurement of her genitalia: women with "hypertrophy" of the clitoris were presumed to be habitual masturbators. The reason, explained Clara Barrus of Middletown State Hospital in New York, in an 1895 report that carefully detailed clitoral abnormalities in 100 patients, was that masturbation stirred blood flow to the external genitalia, which led to the "exaggerated nutrition of these organs" and thus abnormal growth. Since masturbation was viewed as a cause of insanity, some sought to cure it with clitoridectomy, a surgery invented by an English doctor in 1858. However, Barrus found this remedy, which "has been and is still so much in vogue," to be futile:

> It seems to me to be a very reprehensible practice, inasmuch as the worst case of masturbation I have ever seen is that of a young woman who has had clitoridectomy performed. This patient had masturbated, more or less, all her life, and finally, after suffering from several attacks of nymphomania, decided to have the clitoris amputated. The result was not only failure to relieve the nymphomania, but even an increase in its severity, causing a shameless and, almost literally, continuous indulgence in the habit.[11]

While Barrus may have found it objectionable, this surgery did not disappear altogether from American asylums until at least 1950.[12]

Another popular line of investigation focused on endocrine therapies. In the early 1900s, much was being learned about the function of various hormonal glands, leading to speculation that psychotic disorders might be tied to their dysfunction. As a remedy, psychiatrists in the United States and abroad tried injecting the mentally ill with extracts from animals' ovaries, testicles, pituitaries, and thyroids. Extract of sheep thyroid was a particularly popular treatment, having been judged by asylum superintendent William Mabon to have helped nearly 50 percent of his insane patients get better. The extract made the patients quite sick—they grew feverish, lost weight, and their red blood-cell counts declined—but once the treatment ceased, their fevers went away, they gained back weight, and their mental health improved. Mabon, who theorized that the process modified "cell nutrition," reported in 1899 that only one of his healed patients had ever relapsed, suggesting that sheep extract, when it worked, provided a permanent cure.[13]

Other physicians, armed with speculative theories of various sorts, sought to cure their insane patients by injecting toxic chemicals and other foreign substances into their veins, muscles, and cerebrospinal fluid. Injections of metallic salts—manganese, cadmium, and cesium—were tried and found to be worthwhile. The "strychnotonon cure" consisted of a dose of arsotonin, strychnine hydrochloride, and glycerophosphate. One investigator tried the "intraspinal administration of arsenic." Robert Carroll, medical director of Highland Hospitals in Asheville, North Carolina, determined that multiple injections of sterilized horse serum into the spinal fluid, which caused aseptic meningitis, could successfully restore schizophrenics to lucidity. Much like those treated with sheep extract, Carroll's patients had to suffer through physical discomfort for this cure, including backaches, headaches, and vomiting.[14]

Henry Cotton, superintendent at Trenton State Hospital in New Jersey, decided in 1916 that he might be able to cure insanity by removing his patients' teeth. Although Cotton's work eventually led to a medical misadventure of a notable sort, he was a well-trained physician, having studied under the great Swiss psychiatrist Emil Kraepelin and the equally famous Alois Alzheimer, and there was an underlying logic to his seemingly preposterous hypothesis. Bacteria caused many acute illnesses, and various researchers at that time had speculated that "masked" or "hidden" bacterial infections caused chronic ailments like arthritis. Cotton simply applied this general theory to mental illness. He reasoned that teeth were the site of the "masked" infection because there had been scattered reports in the scientific literature, dating back to 1876, of insanity being cured by the removal of infected molars and cuspids. From this initial site of infection, he reasoned, bacteria could spread through the lymph or circulatory systems to the brain, where it "finally causes the death of the patient or, if not that, a condition worse than death—a life of mental darkness."[15] Moreover, when Cotton looked into his patients' mouths, he could always find teeth that were harboring bacteria—evidence, at least to him, that his theory was correct.

He initially removed the infected teeth of fifty chronic patients, only to find that this produced no benefit. Apparently, in chronic patients the deterioration in the brain had already progressed too far, and so Cotton began extracting the teeth of newly admitted patients. This simple procedure, Cotton announced in 1919, cured 25 percent of them. That left 75 percent unimproved, which prompted

Cotton to look for other body regions that might be harboring bacteria. Taking out the patients' tonsils, he said, cured another 25 percent of all new admissions. And if removing their tonsils didn't work, Cotton moved on to their genitourinary and gastrointestinal tracts. This meant surgical removal of a diverse array of body parts: the colon, gall bladder, appendix, fallopian tubes, uterus, ovaries, cervix, and seminal vesicles—they were all targets of Cotton's knife. "We started to literally 'clean' up our patients of all foci of chronic sepsis," he explained.[16]

His "cleaning up" process apparently produced stunning results. Eight-five percent of patients admitted to Trenton State Hospital over a four-year period, he said, had been cured and sent home. Only 3 percent of those who had recovered had ever relapsed; the rest were "earning their living, taking care of families and are normal in every respect."[17] As Cotton was a physician with impeccable credentials, it seemed that at last a true medical breakthrough had been achieved. Burdette Lewis, commissioner of New Jersey's state hospitals, proudly declared that Cotton's "methods of modern medicine, surgery, and dentistry have penetrated the mystery which has enshrouded the subject of insanity for centuries . . . freedom for these patients appears near at hand." Newspapers also sung his praises, as did Adolf Meyer, the "dean" of American psychiatry at that time. Cotton, he said, "appears to have brought out palpable results not attained by any previous or contemporary attack on the grave problem of mental disorder."[18]

However, others who tried his surgeries failed to replicate his good results, and at a 1922 meeting of the American Psychiatric Association, several critics questioned whether Cotton was being "blinded" by his own preconceived ideas. And was it ethical to remove body tissues that appeared to be functioning just fine? "I was taught, and I believe correctly, not to sacrifice a useful part if it could possibly be avoided," one physician said.[19] In 1924, the board for Trenton State Hospital was troubled enough to launch its own investigation. Did Cotton's surgeries work, or not? Meyer was asked to oversee the inquiry, and a review of Cotton's patient records quickly revealed that it was all a sham. Nearly 43 percent of patients who'd undergone Cotton's "thorough treatment" had died. Cotton's "claims and statistics," Meyer confessed to his brother in a letter, "are preposterously out of accord with the facts."[20] Cotton had killed more than 100 patients with his intestinal surgeries alone.*

The first drastic somatic remedy to achieve a more widespread success was deep-sleep therapy, which was popularized by Swiss psychiatrist Jakob Klaesi after World War I. By then, barbiturates—which had been developed by German chemists a decade earlier—were being routinely used in asylums to sedate manic patients, and Klaesi decided to use the drugs to keep patients asleep for days and even weeks on end, hoping that this lengthy rest would restore their nervous systems. He first tried this therapy on a thirty-nine-year-old businesswoman who, following a breakdown, had degenerated to the point where she lay naked in a padded cell. After the prolonged narcosis, Klaesi said, she recovered so fully that her husband marveled at how she was more "industrious, circumspect and tender" than ever before. In the wake of Klaesi's announced success, deep-sleep therapy became quite popular in Europe. Some who tried it claimed that it helped up to 70 percent of their psychotic patients. Enthusiasm for this therapy began to diminish, however, after Swiss psychiatrist Max Muller reported that it had a mortality rate of 6 percent.

Hope was also kindled in the 1920s by the success of malarial fever therapy for general paresis, a type of insanity that occurs in the end-stage of syphilis. This success story had a lengthy history. In 1883, Austrian psychiatrist Julius Wagner-Jauregg noticed that one of his psychotic patients improved during a bout of fever, which led him to wonder whether a high temperature could reliably cure

---

*This scandal was kept from the public, however. Meyer and the hospital board agreed to keep his findings quiet, and although Cotton stopped performing intestinal surgeries, he resumed attacking his patients' teeth, often extracting all of them as this left "no prospect for any further trouble." This form of therapy continued at Trenton State Hospital for twenty years. At Cotton's death in 1933, he was widely eulogized, and Meyer publicly saluted him for having left "an extraordinary record of achievement." Meyer's actions—in essence, he allowed Cotton to continue to perform purposeless, mutilating surgeries rather than expose psychiatry to a black eye—seem inexplicable until it is remembered that he was a member of the advisory board to the American Eugenics Society and had served for a year as president of the Eugenics Research Association. The sordid story of Meyer's coverup was unearthed in 1986 by University of California sociologist Andrew Scull.

schizophrenia. For the next three decades, he occasionally experimented with this idea, using vaccines for tuberculosis and other illnesses to induce potent fevers. He reported some success, but his work failed to draw much attention. Then, during World War I, while working at a clinic in Vienna, he abruptly decided to inject malaria-infected blood into a twenty-seven-year-old man, T. M., ill with paresis. After suffering through nine febrile attacks, T. M. improved so dramatically that soon he was delivering wonderfully coherent lectures on music to other asylum patients.[21]

As a remedy for paresis, malarial fever treatment had an evident biological rationale. Syphilis was known to be an infectious disease. By 1906, the spirochete that causes it had been isolated, and a diagnostic blood test had been developed. The high fevers induced by malaria apparently killed or slowed the spirochete, and thus, at least in some instances, arrested the progress of the disease. In 1927, Wagner-Jauregg was awarded the Nobel Prize in medicine for his work.

Others soon tried fever therapy as a cure for schizophrenia and manic-depressive insanity. Elaborate methods were devised for making patients feverish: hot baths, hot air, electric baths, and infrared and carbon-filament cabinets were all tried. None of this, however, produced impressive results. Mental patients were also deliberately infected with malaria, even though, unlike the paresis patients, they weren't suffering from a known infectious disorder. One physician who tried this, Leland Hinsie at New York State Psychiatric Institute, was remarkably candid about the results: Two of his thirteen patients died, and in several others, "the ill effects were outstanding."[22]

Perhaps the most unusual experiment of all was conducted by two Harvard Medical School physicians, John Talbott and Kenneth Tillotson. Inspired in part by historical accounts of the benefits of extreme cold, they put ten schizophrenic patients between "blankets" cooled by a refrigerant, dropping their body temperatures 10° to 20° Fahrenheit below normal. The patients were kept in this state of "hibernation" for up to three days. Although one of their ten patients died, several others were said to have improved after they were warmed up and returned to consciousness, which in turn led others to toy with this approach. Two Ohio doctors,

Douglas Goldman and Maynard Murray, developed their own version of "refrigeration therapy." They put their mentally ill patients into a cooled cabinet, packed their bodies with ice, and kept them in this refrigerated state for a day or two, with this treatment then periodically repeated. But after three of their sixteen patients died and others suffered a variety of physical complications, they decided, "with a sense of keen disappointment," that refrigeration therapy might not be such a good idea after all.[23]

The Rise of Shock Therapies

Despite the steady pronouncements in medical journals about effective remedies for psychotic disorders, by the early 1930s psychiatry had become ever more discouraged with asylum medicine. Initial claims of success seemed inevitably to be followed by failure. Psychiatrists' sense of therapeutic futility also coincided with society's increasing disregard for the mentally ill. Asylums were being run on impossibly skimpy budgets and were staffed by poorly paid attendants who regularly relied on force to keep the patients in line. Eugenicists had urged that the mentally ill be segregated from society and kept locked up for long periods, and that was precisely what was happening. Asylums in the 1930s were discharging fewer than 15 percent of their patients annually—a rate that was markedly lower than at any time since moral-treatment asylums had been founded in the early 1800s. All of this combined to create the sense that the hospitalized mentally ill were a lost cause and that recovery from severe mental illness was a rare thing. And it was that *pessimism*—along with eugenic attitudes that devalued the mentally ill for who they were—that paved the way for the introduction of shock therapies into asylum medicine.*

---

*The influence of eugenic attitudes on the outcomes of the severely mentally ill is easy to trace. Throughout the 1800s, asylums regularly reported discharging up to 50 percent of their first-episode patients within twelve months of admission. For instance, in 1870, half of the patients at Worcester State Lunatic Asylum had been confined less than a year, and only 14 percent had been confined more than five years. Fairly quick recovery from an acute episode of psychosis was common. Eugenic attitudes toward the mentally ill altered that pattern of recovery. For example, a 1931 long-term study

The first to arrive was insulin-coma therapy. This treatment, pioneered by Viennese psychiatrist Manfred Sakel, was stunning in its boldness. In the late 1920s, while working at private clinic in Berlin, Sakel had discovered that small doses of insulin helped morphine addicts cope with their withdrawal symptoms. On several occasions, however, his patients had lapsed into dangerous hypoglycemic comas, an often fatal complication. But as they returned to consciousness, brought back by an emergency administration of glucose, they appeared changed. Addicts who had been agitated and restless prior to the coma had become tranquil and more responsive. This led Sakel to speculate that if he deliberately put psychotic patients into an insulin coma, something one ordinarily wanted desperately to avoid, they too might awake with altered personalities.

In 1933, Sakel put his audacious idea to the test. After a few trials, he discovered that in order to produce a lasting change, he needed to put patients into deep comas over and over again—twenty, forty, even sixty times over a two-month period. That exhaustive course of therapy, Sakel reported, led to spectacular results: Seventy percent of 100 psychotic patients so treated had been cured, and another 18 percent had notably improved. The cured were "symptom-free," Sakel said, "with full insight into their illness, and with full capacity for return to their former work."[25]

Sakel struggled to explain why the repeated comas benefited schizophrenics. However, it was known that hypoglycemia could cause brain damage, which suggested that trauma itself might be the healing mechanism. Autopsies of people dead from hypoglycemia revealed "widespread degeneration and necrosis of nerve cells," particularly in the cerebral cortex, the brain region responsible for higher intellectual functions.[26] Might the death of brain cells be good for those newly struck by psychosis? Sakel reasoned

---

of 5,164 first-episode patients admitted to New York state hospitals between 1909 and 1911 found that *over the next seventeen years,* only 42 percent were ever discharged (a discharge rate reached in under one year in the 1800s). The remaining 58 percent either died in the hospital or were still confined at the end of that period. But by the 1930s, physicians had forgotten about discharge rates in the 1800s, and contemporary discharge rates convinced them that recovery from severe mental illness was rare.[24]

that the comas selectively killed or silenced "those (brain) cells which are already diseased beyond repair." With the malfunctioning brain cells so killed, the healthy ones could once again become active, leading to a "rebirth" of the patient. His treatment, he said, "is rather a fine microscopic surgery . . . the cure is affected [because it] starves out the diseased cells and permits the dormant ones to come into action in their stead."[27]

Other European investigators reported equally encouraging results. At a meeting in Munsingen, Switzerland, in the spring of 1937, they announced cure rates of 70 percent, 80 percent, and even 90 percent. And this was with schizophrenics, the very class of patients seen as most hopeless. Positive results began rolling in from the United States as well. Joseph Wortis, who had watched Sakel administer insulin therapy at his Vienna clinic, introduced it at Bellevue Hospital in New York City, and he reported recoveries in 67 percent of his patients. In 1938, Benjamin Malzberg from New York State Psychiatric Institute announced positive results from hospitals around the state: Two-thirds of 1,039 schizophrenics treated with insulin-coma therapy had improved, most of them discharged from the hospital, compared to 22 percent of the patients in a control group. A year later, Malzberg was back with an even stronger statement: "The value of the insulin treatment is now definitely established. Every institution that has given it a fair trial has found it to be effective."[28]

American newspapers and magazines quickly celebrated this new medical wonder. The *New York Times* told of patients who had been "returned from hopeless insanity by insulin," explaining that, following the dangerous coma, the "short circuits of the brain vanish, and the normal circuits are once more restored and bring back with them sanity and reality." *Harper's* magazine said that with insulin treatment, aberrant thoughts and feelings are "channeled again into orderly pathways." *Time* explained the therapy's success from a Freudian perspective: As the patient descends into coma, "he shouts and bellows, gives vent to his hidden fears and obsessions, opens his mind wide to listening psychiatrists." *Reader's Digest* was perhaps the most breathless of all. After the repeated comas, it said, "patients act as if a great burden had been lifted from them. They realize that they have been insane, and that the tragedy of that condition is behind them." Its glowing feature was titled "Bedside Miracle."[29]

Psychiatry basked in its newfound glory. Insulin coma, recalled Alexander Gralnick at the American Psychiatric Association's 1943 annual meeting, had opened "new horizons . . . psychiatrists plunged into work and a new measure of hope was added where before mainly despair had prevailed."[30] They did, in fact, now have a therapy that reliably changed the behavior of psychotic patients. They could put newly admitted patients through an intensive course of insulin-coma therapy and regularly discharge the majority back to their families. But it was a therapy that "worked" in a very specific way, one not captured by media tales of bedside miracles.

Insulin, a hormone isolated in 1922, draws sugar from the blood into muscles. The large doses administered to the mentally ill stripped the blood of so much sugar that in the brain, cells would be "starved" of their fuel source and shut down. This cessation of brain activity, Sakel and others observed, occurred in a chronological order that reflected the brain's evolutionary history. The more recently evolved regions of the brain, those that carried out the higher intellectual functions, shut down first, followed by lower brain centers. As patients slid toward coma, they would begin to moan and writhe, such "decebration symptoms . . . indicating that all the higher and most recently developed levels of the brain are more or less out of action," Sakel said.[31] They were in fact now close to death, their brains so depleted of sugar that only the most primitive regions, those controlling basic functions like respiration, were still functioning. Patients would be left in this deep coma for twenty minutes to two hours, then brought back to life with a glucose solution.

As patients emerged from the coma, they would act in needy, infantile ways. They would plaintively ask the surrounding nurses and doctors who they were, often reaching out, like lost children, to hold their nurses' hands or to hang on to their arms. They would suck their thumbs, frequently call out for their mommies, "behaving as if struggling for life."[32] Here is how Sakel described it:

> An adult patient, for example, will say at a particular stage of his awakening that he is six years old. His entire behavior will be childish to the point that the timbre of his voice and his intonation are

absolutely infantile. He misidentifies the examiner and mistakes him for the doctor he had as a child. He asks him in a childish peevish way when he may go to school. He says he has a "tummyache," etc.[33]

This was the behavior that was viewed by Sakel and others as evidence of the patient's return to lucidity. Wortis explained that the treatment "pacified" patients, and that during this awakening period, "patients are free of psychotic symptoms."[34] Another physician said:

[Patients are] childishly simple in mimicry and behavior . . . at this time the patient is by no means any longer out of his mind and beclouded. These infantile reaction-types correspond to the behavior of his primitive personality—it is, so to speak, a regression to an ontogenetically earlier stage, a regression which we might consider in terms of brain pathology to have been called forth by a temporary suppression of the highest levels of mental functioning.[35]

Physicians with Freudian leanings, like Marcus Schatner at Central Islip State Hospital in New York, put this "recovery" into a psychological framework:

The injection of insulin reduces the patient to a helpless baby which predisposes him to a mother transference . . . the patient is mentally sick, his behavior is irrational; this "displeases" the physician and, therefore, the patient is treated with injections of insulin which make him quite sick. In this extremely miserable condition he seeks help from anyone who can give it. Who can give help to a sick person, if not the physician who is constantly on the ward, near the patient and watches over him as over a sick child? He is again in need of a solicitous, tender, loving mother. The physician, whether he realizes it or not, is at present the person who assumes that attitude toward the patient which the patient's mother did when he was a helpless child. The patient in his present condition bestows the love which he once had for his mother, upon the physician. This is nothing else but a mother transference.[36]

This alteration in behavior was also recognized as consistent with brain trauma. One physician compared it to the "behavior of hanged persons after resuscitation, the sick after avalanches . . . the condition which comes on after head injuries, during the progress of uremic coma, after carbon monoxide intoxication and other types of poisoning."[37] However, a single coma did not produce lasting change. Patients would pass through the reawakening state, when they acted like infants, and then their cerebral cortexes would begin to more fully function, and their difficult behavior and fantasies would return. But gradually, if this trauma were repeatedly inflicted, patients would take longer and longer to recover, and their "lucid" periods would become more prolonged. They would now indeed be different. Most notably, they would be less self-conscious. Their own thoughts would interest them less; they would become "detached" from their preoccupations of before. The "emotional charge" that had once fueled their delusions and inner demons would diminish and perhaps even fade away altogether. At times, Sakel acknowledged, the "whole level of (a patient's) personality was lowered." But often, in this new simpler state, they would remain friendlier, more extroverted and "sociable."[38]

Various investigations conducted at the time revealed the nature of the brain damage behind this change. Experiments with cats, dogs, and rabbits showed that insulin comas caused hemorrhages in the brain, destroyed nerve tissue in the cortex, and brought about other "irreversible structural alterations in the central nervous system." Moreover, the greater the number of insulin treatments, "the more severe was the pathology," reported Solomon Katzenelbogen, a psychiatrist at Johns Hopkins Medical School. Autopsies of patients who had died from insulin-coma therapy similarly revealed "areas of cortical devastation." Researchers found evidence of neuronal shrinkage and death, softening of the brain, and general "areas of cellular waste." The pathology often resembled the brain damage that arises from an extended shutoff of oxygen to the brain, leading some to speculate that insulin coma killed cells in this manner as well.[39]

Indeed, this understanding that anoxia, or oxygen depletion to the brain, might be the curative mechanism led to experiments on ways to induce this trauma in a more controlled manner. Harold Himwich, a physician at Albany Medical School in New York, tried doing so by having his patients breathe through a gas mask and then abruptly cutting off the flow of oxygen, replacing it with nitrogen. They would quickly lose consciousness and then be kept in this oxygen-depleted state for a few minutes. Himwich would apply this treatment to his patients three times a week, which led one popular science writer of the day to describe its mechanism of action with an unforgettable turn of phrase: "Schizophrenics don't get well merely by being deprived of oxygen," explained Marie Beynon Ray in *Doctors of the Mind,* which presented Himwich as one of the latest miracle workers in psychiatry. "Occasionally one may recover after [a botched] hanging—but only temporarily. In a few weeks [relapse] . . . But did a lunatic ever get hanged—and hanged—and hanged?"[40]

Insulin-coma therapy remained a common treatment for schizophrenia into the mid-1950s, in spite of periodic reports suggesting that it was doing more harm than good. One problem was its high mortality rate. In 1941, a U.S. Public Health survey found that 5 percent of all state-hospital patients who received the treatment had died from it. But even those who were successfully treated and discharged from the hospital did not fare well over the long term. Patients came back to the mental hospitals in droves, with as many as 80 percent having to be readmitted and most of the rest faring poorly in society. One long-term study found that only 6 percent of insulin-treated patients remained "socially recovered" three years after treatment, which was a markedly worse outcome than for those simply left alone. "It suggests the possibility that the insulin therapy may have retarded or prevented recovery," Ohio investigators sadly concluded in 1950.[41] Other researchers in the mid-1950s echoed this lament, writing of "the insulin myth," which they chalked up to psychiatry's desperate yearning, in the 1930s, for a therapeutic triumph.[42]

In hindsight, it is also evident that many of those harmed by the insulin myth were precisely those patients who would have had the greatest chance of recovering naturally. Sakel had announced early on that the therapy appeared to primarily benefit those who had only recently fallen ill. Moreover, because it was such a hazardous procedure, he wouldn't try it on patients who had other physical ailments, such as kidney disease or a cardiovascular disorder. As Wortis noted, Sakel picked "strong young individuals" with "recent cases." Sakel's results were then confirmed in the United

States by New York asylum physicians who also cherry-picked this healthiest group for the therapy. Even Malzberg admitted that in New York "the insulin-treated patients were undoubtedly a selected group."[43] Not only did this hopelessly bias the initial study results, but it led to the therapy being used, over the years, primarily on physically healthy patients. It turned them into people who, as a result of the brain damage, had little chance to fully recover and live fully human lives.

In the late 1930s, however, insulin-coma therapy "definitely" worked. And it did so in a variety of ways. Patients could be admitted to a hospital, given twenty to sixty comas over a short period, and sent home—an apparent set cure for schizophrenia. Both nurses and physicians found their interactions with the insulin-treated patients much more pleasing as well. Nurses, rather than having to quarrel endlessly with raucous patients, could hover over infantilized, yet sometimes surprisingly cheerful, patients, which made them feel "like I do around small children, sort of motherly." Physicians had the heady experience of performing daily miracles: "I take my insulin therapy patients to the doors of death," said one, "and when they are knocking on the doors, I snatch them back."[44] Patients so treated would spend a great deal of time sleeping between the daily comas, leading to a diminution of noisy, disturbed behavior on the wards, yet another blessing for hospital staff. Hospitals that set up insulin wards could also point to this activity as evidence that they were providing the mentally ill with modern, scientific medicine. All of this made for a medical drama that could be appreciated by many and, further, could evoke public praise.

But for the mentally ill, it represented a new turn in their care. Brain trauma, as a supposed healing therapy, was now part of psychiatry's armamentarium.

The Elixir of Life

For hospitals, the main drawback with insulin-coma therapy was that it was expensive and time consuming. By one estimate, patients treated in this manner received "100 times" the attention from medical staff as did other patients, and this greatly limited its use. In contrast, metrazol convulsive therapy, which was introduced into U.S. asylums shortly after Sakel's insulin treatment arrived, could be administered quickly and easily, with one physician able to treat fifty or more patients in a single morning.

Although hailed as innovative in 1935, when Hungarian Ladislas von Meduna first announced its benefits, metrazol therapy was actually a remedy that could be traced back to the 1700s. European texts from that period tell of using camphor, an extract from the laurel bush, to induce seizures in the mad. Meduna was inspired to revisit this therapy by speculation, which wasn't his alone, that epilepsy and schizophrenia were antagonistic to each other. One disease helped to drive out the other. Epileptics who developed schizophrenia appeared to have fewer seizures, while schizophrenics who suffered seizures saw their psychosis remit. If that was so, Meduna reasoned, perhaps he could deliberately induce epileptic seizures as a remedy for schizophrenia. "With faint hope and trembling desire," he later recalled, "the inexpressible feeling arose in me that perhaps I could use this antagonism, if not for curative purposes, at least to arrest or modify the course of schizophrenia."[45]

After testing various poisons in animal experiments, Meduna settled on camphor as the seizure-inducing drug of choice. On January 23, 1934, he injected it into a catatonic schizophrenic, and soon Meduna, like Klaesi and Sakel, was telling a captivating story of a life reborn. After a series of camphor-induced seizures, L. Z., a thirty-three-year-old man who had been hospitalized for four years, suddenly rose from his bed, alive and lucid, and asked the doctors how long he had been sick. It was a story of a miraculous rebirth, with L. Z. soon sent on his way home. Five other patients treated with camphor also quickly recovered, filling Meduna with a sense of great hope: "I felt elated and I knew I had discovered a new treatment. I felt happy beyond words."

As he honed his treatment, Meduna switched to metrazol, a synthetic preparation of camphor. His tally of successes rapidly grew: Of his first 110 patients, some who had been ill as long as ten years, metrazol-induced convulsions freed half from their psychosis.[46]

Although metrazol treatment quickly spread throughout European and American asylums, it did so under a cloud of great controversy. As other physicians tried it, they published recovery rates that were wildly different. One would find that it helped 70 percent of schizophrenic patients. The next would find that it didn't appear to be an effective treatment for schizophrenia at all but was useful for treating manic-depressive psychosis. Others would find it helped almost no one. Rockland State Hospital in New York announced that it didn't produce a single recovery among 275 psychotic patients, perhaps the poorest reported outcome in all of psychiatric literature to that time.[47] Was it a totally "dreadful" drug, as some doctors argued? Or was it, as one physician wrote, "the elixir of life to a hitherto doomed race?"[48]

A physician's answer to that question depended, in large measure, on subjective values. Metrazol did change a person's behavior and moods, and in fairly predictable ways. Physicians simply varied greatly in their beliefs about whether that change should be deemed an "improvement." Their judgment was also colored by their own emotional response to administering it, as it involved forcing a violent treatment on utterly terrified patients.

Metrazol triggered an explosive seizure. About a minute after the injection, the patient would arch into a convulsion so severe it could fracture bones, tear muscles, and loosen teeth. In 1939, the New York State Psychiatric Institute found that 43 percent of state hospital patients treated with metrazol had suffered spinal fractures. Other complications included fractures of the humerus, femur, pelvic, scapula, and clavicle bones, dislocations of the shoulder and jaw, and broken teeth. Animal studies and autopsies revealed that metrazol-induced seizures caused hemorrhages in various organs, such as the lungs, kidney, and spleen, and in the brain, with the brain trauma leading to "the waste of neurons" in the cerebral cortex.[49] Even Meduna acknowledged that his treatment, much like insulin-coma therapy, made "brutal inroads into the organism."

> We act with both methods as with dynamite, endeavoring to blow asunder the pathological sequences and restore the diseased organism to normal functioning . . . beyond all doubt, from biological and therapeutic points of view, we are undertaking a violent onslaught with either method we choose, because at present nothing less than such a shock to the organism is powerful enough to break the chain of noxious processes that leads to schizophrenia.[50]

As with insulin, metrazol shock therapy needed to be administered multiple times to produce the desired lasting effect. A complete course of treatment might involve twenty, thirty, or forty or more injections of metrazol, which were typically given at a pace of two or three a week. To a certain degree, the trauma so inflicted also produced a change in behavior similar to that seen with insulin. As patients regained consciousness, they would be dazed and disoriented—Meduna described it as a "confused twilight state." Vomiting and nausea were common. Many would beg doctors and nurses not to leave, calling for their mothers, wanting to "be hugged, kissed and petted." Some would masturbate, some would become amorous toward the medical staff, and some would play with their own feces. All of this was seen as evidence of a desired regression to a childish level, of a "loss of control of the higher centres" of intelligence. Moreover, in this traumatized state, many "showed much greater friendliness, accessibility, and willingness to cooperate," which was seen as evidence of their improvement. The hope was that with repeated treatments, such friendly, cooperative behavior would become more permanent.[51]

The lifting in mood experienced by many patients, possibly resulting from the release of stress-fighting hormones like epinephrine, led some physicians to find metrazol therapy particularly useful for manic-depressive psychosis. However, as patients recovered from the brain trauma, they typically slid back into agitated, psychotic states. Relapse with metrazol was even more problematic than with insulin therapy, leading numerous physicians to conclude that "metrazol shock therapy does not seem to produce permanent and lasting recovery."[52]

Metrazol's other shortcoming was that after a first injection, patients would invariably resist another and have to be forcibly treated. Asylum psychiatrists, writing in the *American Journal of Psychiatry* and other medical journals, described how patients would cry, plead that they "didn't want to die," and beg them "in the name of humanity" to stop the injections. Why, some patients would wail, did the hospital want to "kill" them? "Doctor," one woman pitifully asked, "is there no cure for this treatment?" Even military men who had borne "with comparative fortitude and bravery the brunt of enemy action" were said to cower in terror at the prospect of a metrazol injection.

One patient described it as akin to "being roasted alive in a white-hot furnace"; another "as if the skull bones were about to be rent open and the brain on the point of bursting through them." The one theme common to nearly all patients, Katzenelbogen concluded in 1940, was a feeling "of being excessively frightened, tortured, and overwhelmed by fear of impending death."[53]

The patients' terror was so palpable that it led to speculation whether fear, as in the days of old, was the therapeutic agent. Said one doctor:

> No reasonable explanation of the action of hypoglycemic shock or of epileptic fits in the cure of schizophrenia is forthcoming, and I would suggest as a possibility that as with the surprise bath and the swinging bed, the "modus operandi" may be the bringing of the patient into touch with reality through the strong stimulation of the emotion of fear, and that the intense apprehension felt by the patient after an injection of cardiazol [metrazol] and so feared by the patient, may be akin to the apprehension of the patient threatened with the swinging bed. The exponents of the latter pointed out that fear of repetition was an important element in its success.[54]

Advocates of metrazol therapy were naturally eager to distinguish it from the old barbaric shock practices and even conducted studies to prove that fear was not the healing agent. In their search for a scientific explication, many put a Freudian spin on the healing psychology at work. One popular notion, discussed by Chicago psychotherapist Roy Grinker at an American Psychiatric Association meeting in 1942, was that it put the mentally ill through a near-death experience that was strangely liberating. "The patient," Grinker said, "experiences the treatment as a sadistic punishing attack which satisfies his unconscious sense of guilt."[55] Abram Bennett, a psychiatrist at the University of Nebraska, suggested that a mental patient, by undergoing "the painful convulsive therapy," has "proved himself willing to take punishment. His conscience is then freed, and he can allow himself to start life over again free from the compulsive pangs of conscience."[56]

As can be seen by the physicians' comments, metrazol created a new emotional tenor within asylum medicine. Physicians may have reasoned that terror, punishment, and physical pain were good for the mentally ill, but the mentally ill, unschooled in Freudian theories, saw it quite less abstractly. They now perceived themselves as confined in hospitals where doctors, rather than trying to comfort them, physically assaulted them in the most awful way. Doctors, in their eyes, became their *torturers*. Hospitals became places of torment. This was the beginning of a profound rift in the doctor-patient relationship in American psychiatry, one that put the severely mentally ill ever more at odds with society.

Even though studies didn't provide evidence of any long-term benefit, metrazol quickly became a staple of American medicine, with 70 percent of the nation's hospitals using it by 1939. From 1936 to 1941, nearly 37,000 mentally ill patients underwent this treatment, which meant that they received multiple injections of the drug. "Brain-damaging therapeutics"—a term coined in 1941 by a proponent of such treatments—were now being regularly administered to the hospitalized mentally ill, and being done so against their will.[57]

The Benefits of Amnesia

The widespread use of metrazol provided psychiatry, as a discipline, with reason for further optimism and confidence. Asylum doctors now had two treatments that could reliably induce behavioral change in their patients. A consensus emerged that insulin coma was the preferred therapy for schizophrenia, with metrazol best for manic-depressive disorders. At times the two methods would be combined into a single treatment, a patient first placed into a deep coma with insulin and then given a metrazol injection to induce seizures. "All of this has had a tremendously invigorating effect on the whole field of psychiatry," remarked A. Warren Stearns, dean of Tufts Medical School, in 1939. "Whereas one often sent patients to state hospitals solely for care, it has now become possible to think in terms of treatment."[58] Psychiatry, as it moved forward, could hope to build on these two therapeutic successes.

Electroshock, the invention of Italian psychiatrist Ugo Cerletti, did just that. Cerletti, head of the psychiatry department at the University of Rome, had been deeply impressed by both Sakel's and

Meduna's triumphs, and his own research suggested a way to improve on metrazol therapy. For years, as part of his studies of epilepsy, he had been using electricity to induce convulsions in dogs. Other scientists, in fact, had been using electricity to induce convulsions in animals since 1870. If this technique could be adapted to humans, it would provide a much more reliable convulsive method. The problem was making it safe. In his dog experiments—Cerletti would place one electrode in the dog's mouth and one in the anus—half of the animals died from cardiac arrest. The United States even regularly killed its criminals with jolts of electricity, which gave Cerletti pause. "The idea of submitting man to convulsant electric discharges," he later admitted, was considered "barbaric and dangerous; in everyone's mind was the spectre of the electric chair."[59]

As a first step in this research, Cerletti's assistant Lucio Bini studied the damage to the nervous system produced by electricity-induced convulsions in dogs. He found that it led to "acute injury to the nerve cells," particularly in the "deeper layers of the cerebral cortex." But Bini did not see this damage necessarily as a negative. It was, he noted, evidence that "anatomical changes can be induced." Insulin coma also produced "severe and irreversible alterations in the nervous system," and those "very alterations may be responsible for the favorable transformation of the morbid psychic picture of schizophrenia. For this reason, we feel that we are justified in continuing our experiments."[60]

The eureka moment for Cerletti, however, came in a much more offbeat venue—a local slaughterhouse. Cerletti had gone there expecting to observe how pigs were killed with electroshock, only to discover that the slaughterhouse simply stunned the pigs with electric jolts to the head, as this made it easier for butchers to stab and bleed the animals. The key to using electricity to induce seizures in humans, Cerletti realized, was to apply it directly to the head, rather than running the current through the body. After testing this premise in animal experiments, he said, "I felt we could venture to experiment on man, and I instructed my assistants to be on the alert for the selection of a suitable subject."[61]

The suitable subject turned out to be a thirty-nine-year-old disoriented vagrant rounded up at the railroad station by Rome police and sent to Cerletti's clinic for observation. "S. E.," as Cerletti called him, was from Milan, with no family in Rome. Later, Cerletti would learn that S. E. had been previously treated with metrazol, but he knew little of S. E.'s past when, in early April 1938, he conducted his bold experiment. At first, it went badly. Neither of the initial two jolts of electricity, at 80 and 90 volts, successfully knocked out S. E.—he even began singing after the second. Should the voltage be increased? As Cerletti and his team discussed what to do—his assistants thought a higher dose would be lethal—S. E. suddenly sat up and protested: *"Non una seconda! Mortifera!"* ("Not a second! It will kill me!") With those words ringing in his ears, Cerletti, intent on not yielding "to a superstitious notion," upped the jolt to 110 volts, which quickly sent S. E. into a seizure. Soon Cerletti trumpeted his achievement: "That we can cause epileptic attacks in humans by means of electrical currents, without any danger, seems to be an accepted fact."[62]

Electroshock, which was introduced into U.S. hospitals in 1940, was not seen as a radical new therapy. As Cerletti had suggested, his achievement had simply been to develop a better method for inducing convulsions. Electricity was quick, easy, reliable, and cheap—all attributes that rapidly made it popular in asylum medicine. Yet, as soon became clear, electroshock also advanced "brain-damaging therapeutics" a step further. In comparison with metrazol, it produced a more profound, lasting trauma. Sakel, who thought the trauma too extreme, pinpointed the difference from his own insulin treatment: "In the amnesia caused by all electric shocks, the level of the whole intellect is lowered . . . the stronger the amnesia, the more severe the underlying brain cell damage must be."[63]

Indeed, asylum medicine was now pitching headlong down a very peculiar therapeutic path. Was the change effected by brain trauma a good or a bad thing? How one answered that question depended in great part on one's beliefs about the potential for the severely mentally ill to recover and whether there was much to value in them as they were. Criticism of the shock therapies, which came primarily from Freudians, was memorably articulated in 1940 by Harry Stack Sullivan, a leading psychoanalyst:

> These sundry procedures, to my way of thinking, produce "beneficial" results by reducing the patient's capacity for being human. The

> philosophy is something to the effect that it is better to be a contented imbecile than a schizophrenic. If it were not for the fact that schizophrenics can and do recover; and that some extraordinarily gifted and, therefore, socially significant people suffer schizophrenic episodes, I would not feel so bitter about the therapeutic situation in general and the decortication treatments in particular.[64]

Electroshock, the newest "decortication" treatment in asylum medicine, worked in a predictable manner. With the electrodes placed at the temples, the jolt of electricity passed through the temporal lobes and other brain regions for processing memory. As patients spasmed into convulsions, they immediately lost consciousness, the brain waves in the cerebral cortex falling silent. "A generalized convulsion," explained Nolan Lewis, of the New York State Psychiatric Institute, in 1942, "leaves a human being in a state in which all that is called the personality has been extinguished."[65] When patients came to, they would be dazed, often not quite sure of who they were, and at times sick with nausea and headaches. Chicago psychiatrist Victor Gonda noted that patients, in this stunned state, "have a friendly expression and will return the physician's smile."[66]

Even after a single treatment, it would take weeks for a patient's brain-wave activity, as measured by an electroencephalograph, to return to normal. During this period, patients frequently exhibited evidence of "organic neurasthenia," observed Lothar Kalinowsky, who established an electroshock program at New York State Psychiatric Institute in 1940. "All intellectual functions, grasp as well as memory and critical faculty, are impaired." Patients remained fatigued, "disoriented in space and time . . . associations become poor."[67] They also acted in submissive, helpless ways, a change in behavior that made crowded wards easier to manage.

Early on, it was recognized that the dulling of the intellect was the therapeutic mechanism at work. Psychosis remitted because the patient was stripped of the higher cognitive processes and emotions that give rise to fantasies, delusions, and paranoia. As one physician said, speaking of brain-damaging therapeutics: "The greater the damage, the more likely the remission of psychotic symptoms."[68] Said another: "The symptoms become less marked at the same time as a general lowering of the mental level occurs."[69]

Research even directly linked the slowing of brain wave activity to diminishment of "hallucinatory activity."

The memory loss caused by electroshock was also seen as helpful to the mentally ill. Patients, physicians noted, could no longer recall events that had previously caused them so much anguish. "The mechanism of improvement and recovery seems to be to knock out the brain and reduce the higher activities, to impair the memory, and thus the newer acquisition of the mind, namely the pathological state, is forgotten," explained Boston psychiatrist Abraham Myerson, speaking at the American Psychiatric Association's annual meeting in 1943.[70]

As quickly became evident, however, electroshock's "curative" benefits dissipated with time. When patients recovered from the trauma, their mental illness often returned. "That relapses will come, that in many cases the psychosis remanifests itself as the brain recovers from its temporary injury is, unfortunately, true," Myerson admitted. "But the airplane has flown even if shortly it has crashed." Given that problem, logic suggested a perverse next step. If the remission of symptoms were the desired outcome, and if symptoms returned as patients recovered from the head injury, then perhaps electroshock should be repeated numerous times, or even on a daily basis, so that the patient became more deeply impaired. In his 1950 textbook *Shock Treatment,* Kalinowsky dubbed this approach "confusional" treatment. "Physicians who treat their patients to the point of complete disorientation are highly satisfied with the value of ECT [electroshock] in schizophrenia," he noted.[71] Bennett, echoing Kalinowsky's arguments, advised that at times a patient needed to be shocked multiple times to reach "the proper degree of therapeutic confusion."[72] Such guidance led Rochester State Hospital in New York to report in 1945 that its mental patients were being shocked three times weekly, as "this regime has to some extent increased and maintained [their] confusion." The patients were said to be "more amused than alarmed by this circumstance."[73]

One woman so treated was seventeen-year-old Jonika Upton. Her family committed her to Nazareth Sanatorium in Albuquerque, New Mexico, on January 18, 1959, upset that she had run off to Santa Cruz, California, several weeks earlier with a twenty-two-year-old artist boyfriend. Her family was also alarmed that she'd previously had a boyfriend whom they suspected of being "homosexual," that she had developed peculiar speech mannerisms, and that she often "walked about carrying 'Proust' under her arm." Her admissions record described her as "alert and cooperative but [she] makes it plain that she doesn't like it here."[74]

Over the next three months, Upton was shocked sixty-two times. During this course of treatment, her doctors regularly complained about her slow progress: "Frankly," her supervising physician wrote on March 24, "she has not become nearly as foggy as we might wish under such intensive treatment but, of course, there is considerable confusion and general dilapidation of thought." Two weeks later, the doctor's lament was the same: "We are not really satisfied with her reactions to intensive treatment up to the time. Under this type of treatment a patient usually shows a great deal more fogging and general confusion than she has." But by the end of April, Jonika Upton had finally deteriorated to the desired "confusional" point. She was incontinent, walked around naked, and was no longer certain whether her father was dead or alive. A few days later, she was seen as ready for discharge. She was handed back over to her parents, whom she "did not seem to recognize," the nurses observed. However, her symptoms had indeed remitted. Her memory of her boyfriend had been erased, and certainly she was no longer carrying Proust under her arm. Upton's physician chalked her up as a therapeutic success, writing, on the day of her discharge, to a fellow doctor: "She showed marked changes in her thinking and feeling and I believe that she has developed some insight."*

---

*Intensive electroshock was also tried on "schizophrenic" children. Starting in 1942, physicians at New York City's Bellevue Hospital enrolled 98 children, ages four to eleven, in a study that involved shocking them twice daily for twenty days in a row. The Bellevue doctors reported that the treatment successfully made the children "less exciteable, less withdrawn, and less anxious." A few years later, researchers at a different hospital who followed up on the children found that a number had become particularly violent and disturbed. A ten-year-old boy wanted to "kill the physicians who had treated him"; he eventually assaulted his mother for "consenting to this form of treatment" and then "attempted to jump out an apartment window." A nine-year-old boy attempted to hang himself, explaining that he was afraid of being shocked again. Despite this follow-up study, Bellevue Hospital's Lauretta Bender wrote in 1955 that she had successfully put a toddler, not yet three years old, through the twenty-shock ritual.[75]

By the time Upton was treated, researchers had better identified the basic nature of the trauma inflicted by electroshock. Max Fink, at Hillside Hospital in Long Island, who was a proponent of the treatment, had shown that electroshock, as a single event, produced changes very similar to "severe head trauma." The alterations in brain-wave activity were the same, and both produced similar biochemical changes in the spinal fluid. In fact, electroshock did not produce changes similar to epileptic seizures but rather induced changes similar to a concussive head injury. The similarity was such that "convulsive therapy provides an excellent experimental method for studies of craniocerebral trauma," Fink concluded.[76]

What remained controversial was whether such trauma led to permanent brain damage. Although there was much debate on this question, a number of studies had turned up evidence that it did, making intensive treatment that much more problematic. At autopsy and in animal experiments, researchers had found that electroshock could cause hemorrhages in the brain, particularly in the cerebral cortex. "Areas of [cortical] devastation" were found in one patient who died. "Increased neuronal degeneration and gliosis" were reported in 1946. Various investigators announced that repeated electroshock treatments could lead to "permanent impairment of behavioral efficacy and learning capacity," "lower cognitive functioning," "extended memory loss," and a "restriction in intuition and imagination and inventiveness." Leon Salzman, from St. Elizabeth's Hospital in Washington, D.C., noted in 1947 that "most workers agree that the larger the number of shocks the greater damage produced." One year later, a study of schizophrenics shocked more than 100 times found that in some, their inner lives were "apparently barren," their "ego function . . . extremely reduced," and their perception of reality extremely impaired.[77]

Repetitious craniocerebral trauma, as an agent of behavioral change, apparently exacted a high cost.

No Consent Necessary

Asylum doctors, when writing in their medical journals, were fairly candid about the mechanism at work in electroshock, admitting it was a form of brain trauma. But that is not how it was presented to the public. Instead, the popular message was that it was safe, effective, and painless, and that any memory loss was temporary. Journalists writing exposés about the horrible conditions inside state hospitals in the 1940s even held it up as an advanced scientific treatment that should be offered to all. "Patients get a break at Brooklyn, both on the humane and medical end," wrote Albert Deutsch, in *Shame of the States*, pointing to Brooklyn State Hospital as a model for reform. "Virtually every patient who is admitted gets an early chance at shock therapy."[78]

Behind this public facade of humanitarian care, however, a remarkable episode in American medicine was unfolding: Much as patients had resisted metrazol injections, so most resisted electroshock.

Until muscle-paralyzing agents were introduced, the physical trauma from electroshock was much the same as it was for metrazol injection: Up to 40 percent of patients suffered bone fractures with electroshock. This problem was lessened after Bennett reported in 1940 that the drug curare could be used to temporarily paralyze patients, preventing the wild thrashing that could break bones. But such paralyzing drugs were not always given, and even eliminating the bodily trauma didn't eliminate patients' fears. After experiencing shock a few times, Kalinowsky said, some patients "make senseless attempts to escape, trying to go through windows and disregarding injuries." They "tremble," "sweat profusely," and make "impassioned verbal pleas for help," reported Harvard University's Thelma Alper. Electroshock, patients told their doctors, was like "having a bomb fall on you," "being in a fire and getting all burned up," and "getting a crack in the puss." Researchers reported that the mentally ill regularly viewed the treatment as a "punishment" and the doctors who administered it as "cruel and heartless."[79]

That is how doctors, in their more candid moments, reported on their patients' reactions. In their own writings, patients regularly described electroshock in even stronger terms as a horrible assault. In her 1964 memoir, *The White Shirts*, Ellen Field told of the great terror it evoked:

> People tend to underrate the physical damage of anticipating shock. At any rate, they think of it as purely a mental fear. This is so false. The truth is that electric shock is physical torture of an

> extreme type . . . the fear is intensely physical . . . The heart and solar plexus churn and give off waves of—I don't know the word for it. It hasn't the remotest resemblance to anything I've ever felt before or since. Soldiers just before a battle probably experience this same abdominal sensation. It is the instinct of a living organism to fear annihilation.[80]

Sylvia Plath, in *The Bell Jar,* described how it led to both physical and emotional trauma:

> Doctor Gordon was fitting two metal plates on either side of my head. He buckled them into place with a strap that dented my forehead, and gave me a wire to bite. I shut my eyes. There was a brief silence, like an indrawn breath. Then something bent down, and took hold of me and shook me like the end of the world. Whee-ee-ee-ee-ee, it shrilled, through an air crackling with blue light, and with each flash a great jolt drubbed me till I thought my bones would break and the sap fly out of me like a split plant. I wondered what terrible thing it was that I had done.[81]

Dorothy Washburn Dundas, a young woman when she was shocked, recounted in *Beyond Bedlam* a similar story: "My arms and legs were held down. Each time, I expected I would die. I did not feel the current running through me. I did wake up with a violent headache and nausea every time. My mind was blurred. And I permanently lost eight months of my memory for events preceding the shock treatments. I also lost my self-esteem. I had been beaten down."[82] Others described hospital scenes of patients being dragged screaming into the shock rooms. "Believe me when I say that they don't care how they get you there," Donna Allison wrote, in a letter to the editor of a Los Angeles paper. "If a patient resists, they will also choke him until he passes out, and lay him on his bed until he comes to, and then give him treatment. I have also had that happen to me."[83]

Faced with such resistance, American physicians and hospitals simply asserted the right to shock patients without their consent. Historian Joel Braslow, in his review of California patient records, found that only 22 percent of shocked patients had agreed to the treatment, and this was so even though physicians regularly told their hospitalized patients that electroshock was safe and painless.[84] "We prefer to explain as little as possible to the uninformed patient," Bennett explained in 1949. Shock, he said, should be described to patients as "sleep induced by electricity," the patients assured that "there is no pain or discomfort."[85] Other leading electroshock doctors, like David Impastato at Bellevue Hospital in New York City, argued that the mentally ill shouldn't even be told that they were going to be shocked: "Most patients associate EST with severe insanity and if it is suggested, they will refuse it claiming they are not insane and do not need the treatment . . . I recommend that patients be kept in ignorance of the planned treatment."[86] Such forced treatment might not even be remembered, shock advocates reasoned, as patients often had "complete amnesia for the whole treatment."[87]

Such thinking reflected, of course, societal views about the rights—or non-rights—of the mentally ill. By any standard, electroshock was a profound event. Psychiatrists saw it as a method, in Fink's words, to produce "an alteration in brain function."[88] It was designed to change the mentally ill in a pronounced way. The treatment might make their psychotic symptoms and depression disappear, but such relief would come at the cost of their ability to think, feel, and remember, at least for a period of time. Yet the prevailing opinion among America's leading electroshock doctors in the 1940s and 1950s was that in the confines of mental hospitals, they had the right to administer such treatment without the patient's consent, or even over the patient's screaming protests—a position that, if it had been applied to criminals in prison, would have been seen as the grossest form of abuse. Indeed, after World War II ended, when the United States and its allies attended to judging Nazi crimes, the International Red Cross determined that prisoners in concentration camps who had been electroshocked should be compensated for having suffered "pseudomedical" experiments against their will. As some of the shocked prisoners were later killed, "the electroshock treatments could be seen as a prelude to the gas chamber," noted historian Robert Lifton.[89] But in the United States, forced electroshock remained a common practice for more than two decades, with easily more than 1 million Americans subjected to it.

Like so many somatic remedies of earlier periods, electroshock was also used to frighten, control, and punish patients. Braslow found that in California, asylum physicians regularly prescribed electroshock for those who were "fighting," "restless," "noisy," "quarrelsome," "stubborn," and "obstinate"—the treatment made such patients "quieter" and "not so aggressive."[90] Other scholars, writing in medical journals, reported how physicians and hospital staff chose to shock patients they most disliked. One physician told of using it to give women a "mental spanking." An attendant confessed: "Holding them down and giving them the treatment, it reminded me of killing hogs, throwing them down in the pen and cutting their throats." Hospital physicians spoke of giving misbehaving patients "a double-sized course" of electricity.[91]

Many hospitals used electroshock to quiet the wards and set up schedules for mass shocking of their patients. "Patients could look up the row of beds," Dr. Williard Pennell told the *San Francisco Chronicle,* "and see other patients going into epileptic seizures, one by one, as the psychiatrists moved down the row. They knew their turn was coming."[92] Bellevue Hospital in New York touted the use of electroshock as a "sedative" for acutely disturbed patients, shocking them twice a day, which left the "wards quieter and more acceptable to all patients."[93] And at Georgia's Millidgeville Asylum, where 3,000 patients a year were being shocked in the early 1950s, nurses and attendants kept patients in line by threatening patients with a healthy dose of a "Georgia Power cocktail." Superintendent T. G. Peacock informed his attendants: "I want to make it clear that it is hospital policy to use shock treatment to insure good citizenship."[94]

Such was the way electroshock was commonly used in many U.S. mental hospitals in the 1940s and 1950s. Head trauma, if truth be told, had replaced the whip of old for controlling the mentally ill.

5

BRAIN DAMAGE AS MIRACLE THERAPY

It has been said that if we don't think correctly, it is because we haven't "brains enough." Maybe it will be shown that a mentally ill patient can think more clearly and constructively with less brain in actual operation.

—Walter Freeman, 1941[1]

INSULIN COMA, METRAZOL, and electroshock had all appeared in asylum medicine within the space of a few years, and they all "worked" in a similar manner. They all dimmed brain function. Yet they were crude methods for achieving this effect. With these three methods, there was no precise control over the region of the brain that was disabled, nor was there control over the degree to which the brain was traumatized. The approach, said one physician, seemed akin to "trying to right a watch with a hammer."[2] However, during this same period, there was one other therapy that was introduced into asylums which was not so imprecise, and it was this last therapy that, in the 1940s, became psychiatry's crowning achievement. Newspapers and magazines wrote glowing articles about this "miracle" of modern medicine, and, in

1949, fourteen years after its introduction, its inventor, Portuguese neurologist Egas Moniz, was awarded a Nobel Prize.

That therapy, of course, was prefrontal lobotomy.

Inspiration . . . Or a Clear Warning?

The frontal lobes, which are surgically disabled during prefrontal lobotomy, are the most distinguishing feature of the human brain. Put an ape brain and a Homo sapiens brain side by side, and one difference immediately jumps out—the frontal lobes in the human brain are much more pronounced. This distinguishing anatomy, so visible at autopsy, led philosophers as far back as the Greeks to speculate that the frontal lobes were the center for higher forms of human intelligence. In 1861, long before Moniz picked up his drill, the great French neurologist Pierre Paul Broca pointed to the frontal lobes as the brain region that gives humankind its most noble powers:

> The majesty of the human is owing to the superior faculties which do not exist or are very rudimentary in all other animals; judgment, comparison, reflection, invention and above all the faculty of abstraction, exist in man only. The whole of these higher faculties constitute the intellect, or properly called, understanding, and it is this part of the cerebral functions that we place in the anterior lobes of the brain.[3]

Scientific investigations into frontal-lobe function had been jump-started a few years earlier by the remarkable case of Phineas Gage. Gage, a twenty-five-year-old Vermont railroad worker, was preparing a hole for blasting powder when an explosion drove a 3.5-foot iron rod into his left cheek and through his frontal lobes. Incredibly, he survived the accident and lived another fifteen years. But the injury dramatically changed him. Before, others had admired him as energetic, shrewd, and persistent. He was said to have a well-balanced mind. After his accident, he became ill mannered, stubborn, and rude. He couldn't carry out any plans. He seemed to have the mind of a spoiled child. He had changed so radically that his friends concluded that he was "no longer Gage."

Over the next eighty years, animal research revealed similar insights about the importance of the frontal lobes. In 1871, England's David Ferrier reported that destroying this brain region in monkeys and apes markedly reduced their intelligence. The animals, selected for their "intelligent character," became "apathetic or dull or dozed off to sleep, responding only to the sensations or impressions of the moment."[4] Their listlessness was periodically interrupted by purposeless wanderings. Italian neurologist Leonardo Bianchi, who conducted lobotomy experiments in dogs, foxes, and monkeys, concluded in 1922 that the human intelligence responsible for creating civilization could be found in the frontal lobes.

In the 1930s, Carlyle Jacobsen at Yale University conducted studies with two chimps, Becky and Lucy, that highlighted the importance of the frontal lobes for problem solving. He tested this skill through a simple experiment. Each chimp would be placed into a chamber and allowed to watch while food was placed beneath one of two cups. A blind would be lowered, hiding the cups from view, and then, five minutes later, the blind would be raised and the chimp would be given an opportunity to get the food by picking the right cup. After their frontal lobes were removed, Becky and Lucy lost their ability to solve this simple test. The frontal lobes, Jacobsen concluded, were responsible for an organism's adjustment to its environment. This region of the brain synthesized information, including memories formed from recent events, and it was this process that produced intelligent action.[5]

By this time, numerous clinical reports had also documented the effects of severe head wounds. After World War I, Gage's story was no longer such an anomaly. Clinicians reported that people with frontal-lobe injuries became childish and apathetic, lost their capacity to plan ahead, and could not make sound judgments. Similarly, cancer patients who had frontal-lobe operations because of brain tumors were said to act in puerile ways, to lack initiative and will, and to display emotions that seemed flattened or out of sync with events. Frontal-lobe injuries led to a recognizable syndrome, dubbed "Witzelsucht," that was characterized by childish behavior.

None of this intellectual loss and behavioral deterioration following frontal-lobe injury was surprising. If anything, physicians voiced surprise that the intellectual deficits weren't greater. It was remarkable that Gage, who'd had a rod go completely through the front of his brain, could function as well as he did. Ferrier had noted that the extent of the intellectual deficit in his lobotomized monkeys was not immediately evident, but rather became apparent only after some time. People with frontal-lobe injuries were even found to do fairly well on standardized intelligence tests.

Indeed, frontal-lobe injury appeared to produce an odd mixture. The pronounced emotional and problem-solving deficits were accompanied by the retention of a certain mechanical intelligence. Such was the case with Joe A., a New York City stockbroker who developed a brain tumor at age thirty-nine. After Johns Hopkins neurosurgeon Walter Dandy removed the tumor in an operation that caused extensive damage in the prefrontal region of Joe's brain, Joe became a profoundly different person. In some ways, he functioned remarkably well. He could still play checkers, his memory seemed unimpaired, and he understood what had happened to him. At times, he could socialize well. On one occasion, a group of visiting neurologists spent an hour with him and failed to notice anything unusual. But like Gage, Joe was a changed person. He couldn't focus his attention any more, he lacked motivation to go back to work, he couldn't plan daily activities, and he often behaved in emotionally inappropriate ways. He was easily irritated, constantly frustrated, spoke harshly of others, and became a hopeless braggart. He would see boys playing baseball and blurt out that he would soon become a professional ballplayer, as he was a better hitter than anyone. On IQ tests he now scored below ninety, and he could do well only with familiar material. His capacity to learn had disappeared.

Together, the animal studies and clinical reports of head injuries seemingly pointed to a stark conclusion: Destroying tissue in this brain region would cause many intellectual and emotional deficits. The person would likely become more apathetic, lack the ability to plan ahead, be unable to solve problems, and behave in puerile, emotionally inappropriate ways. Witzelsucht was not a kind fate. Yet in 1935, Portuguese neurologist Egas Moniz saw something encouraging in these reports. He found reason to believe that inflicting injury on the frontal lobes could prove beneficial to the mentally ill.

Planting the Seed

The idea of drilling holes into the brains of the mentally ill to cure them was not, in the 1930s, new to psychiatry. As far back as the twelfth century, surgeons had reasoned that trepanning, which involved cutting holes in the scalp, allowed demons to escape from a poor lunatic's brain. In 1888, Gottlieb Burckhardt, director of an asylum in Prefarigier, Switzerland, had removed part of his patients' cerebral cortex to quiet their hallucinations. "If we could remove these exciting impulses from the brain mechanism," he wrote, "the patient might be transformed from a disturbed to a quiet dement."[6] Although one of his six patients died, Burckhardt concluded that it did make the others more peaceful. Twenty years later, a Russian surgeon, Ludwig Puusepp, tried to cure three depressed patients by cutting into their frontal lobes. But when he didn't find it particularly helpful, the notion was pushed to the background of psychiatric research.

Moniz resurrected it at a very telling time in his career. In 1935, Moniz was sixty-one years old. He'd led a colorful, prosperous life, but he had never realized his grandest dreams. As a young man, newly graduated from medical school, he'd thrown himself into political struggles to replace Portugal's monarchy with a democratic government, a struggle that twice landed him in jail. After a new government was established in 1910, he was elected to the Portuguese Parliament and served as ambassador to Spain. Wherever he went, he lived the good life; the parties that he and his wife gave were known for their elegance, style, and good food. But in 1926, it all came tumbling down when the Portuguese government was overthrown in a military coup. Disappointed, even bitter, over the loss of his beloved democracy, he turned his attention full time to medicine and his neurology practice. He'd long juggled his life as an academic physician, on the faculty at the University of Lisbon, with that of his life in politics, and he set his sights on making a lasting contribution to medicine. "I was always dominated by the desire to accomplish something new in the scientific world," he recalled in his memoirs. "Persistence, which depends more on willpower than intelligence, can overcome difficulties which seem at first unconquerable."[7]

Moniz quickly found himself on the verge of the fame he so avidly sought. In 1928, he was nominated for the Nobel Prize in medicine for inventing a technique for taking X-rays of cerebral arteries. He didn't win, though, and he found himself obsessed with the prize. Over the next few years, he actively campaigned to be renominated for the award, at times wielding his pen to disparage others working on similar blood-imaging techniques, fearful that their achievements might diminish his own. Although he was nominated for the Nobel Prize again in 1933, once more the award went to another scientist, and it seemed certain now that the top honor would never be his. That is, until he traveled in August 1935 to London to attend the Second International Congress in Neurology.

That year, the conference featured an all-day symposium on the frontal lobes. A number of speakers presented their latest research on this region of the brain. American neurologist Richard Brickner provided an update on Joe A., his tumor patient. Jacobsen detailed his experiments with the chimps Lucy and Becky. Although fascinating, the reports led to a sobering conclusion. "There is little doubt," wrote George Washington University neurologist Walter Freeman, "but that the audience was impressed by the seriously harmful effects of injury to the frontal lobes and came away from the symposium reinforced in their idea that here was the seat of the personality and that any damage to the frontal lobes would inevitably be followed by grave repercussions upon the whole personality."[8]

Moniz, however, plucked from the presentations a different message. The reports by Jacobsen and Brickner had set his mind churning. Jacobsen, after detailing the cognitive deficits in the chimps following lobotomy, had noted that the surgery also produced a marked emotional change in one of them, Becky. Before the surgery, she had typically reacted angrily when she failed to pick the right cup in the food experiment. She would roll on the floor, defecate, or fly into a rage. But after the surgery, nothing seemed to bother her. If she failed to solve a problem, she would no longer throw an emotional tantrum. It was as though she had joined a "happiness cult" or placed her "burdens on the Lord," Jacobsen said.[9] Brickner's account of Joe A. had made even a deeper impression on Moniz. Although Joe may have changed after his frontal lobes were damaged, apparently he could still be sociable and converse in a relatively normal way. All of which set Moniz to thinking: Could the same be said of time spent with the mad, the emotionally distressed? Who didn't immediately notice their illness? Joe A., Moniz figured, functioned at a much higher level than those ill with schizophrenia or severe depression. What if he deliberately injured both frontal lobes of the mentally ill in order to cure them? After all, Joe A. could "still understand simple elements of intellectual material," he reasoned. "Even after the extirpation of the two frontal lobes, there remains a psychic life which, although deficient, is nevertheless appreciably better than that of the majority of the insane."[10]

Moniz, who prided himself on being a man of science, quickly came up with a neurological explanation for why such surgery would cure the mentally ill. Thoughts and ideas, he reasoned, were stored in groups of connected cells in the brain. Schizophrenia and emotional disorders resulted from pathological thoughts becoming "fixed" in such "celluloconnective systems," particularly in the frontal lobes. "In accordance with the theory we have just developed," he said, "one conclusion is derived: to cure these patients we must destroy the more or less fixed arrangements of cellular connections that exist in the brain."[11]

Three months after returning from London, Moniz chose a sixty-three-year-old woman from a local asylum to be his first patient. He knew his reputation was at stake. Should the operation fail, he would be condemned for his "audacity." The woman, a former prostitute, was paranoid, heard voices, and suffered bouts of crippling anxiety. Moniz's assistant, Almeida Lima, performed the surgery: He drilled holes into her skull, used a syringe to squirt absolute alcohol onto the exposed white fibers, which killed the tissue through dehydration, and then sewed her back up. The whole operation took about thirty minutes. Just hours later, she was able to respond to simple questions, and within a couple of days, she was returned to the asylum. A young psychiatrist there soon reported that the woman had remained calm, with her "conscience, intelligence, and behavior intact," leading Moniz—who'd hardly seen her after the operation—to happily pronounce her "cured."[12]

Within three months, Moniz and Lima operated on twenty mentally ill patients. During this initial round of experimentation, they continually increased the scope of brain damage. The greater the damage, it appeared, the better the results. More holes were drilled, more nerve fibers destroyed. Starting with the eighth patient, Lima began using a thin picklike instrument with a wire loop, called a leucotome, to cut the nerve fibers in the frontal lobes. Each cutting of nerve tissue within was counted as a single "coring"; by the twentieth patient, Lima was taking six such corings from each side of the brain. They also concluded that while the surgery didn't appear to help schizophrenics, it did reliably make those ill with manic depression less emotional. That was all the change that Moniz needed to see. In the spring of 1936, he announced his stunning success: Seven of the twenty patients had been cured. Seven others had significantly improved. The other six were unchanged. "The intervention is harmless," Moniz concluded. "None of the patients became worse after the operation."[13]

Moniz had achieved the triumph he'd long sought. All his fears could now be put to rest. He was certain that his surgery marked "a great step forward." Within a short period, he churned out a 248-page monograph, *Tentatives opératoires dans le traitement de certaines psychoses,* and published his results in eleven medical journals in six countries.* Reviewers in several countries found his lengthy monograph impressive, and none was more enthusiastic than an American, Walter Freeman. Writing in the *Archives of Neurology,* he suggested that, if anything, Moniz had been too "conservative" in his declarations of success. From Freeman's perspective, Moniz's count of seven cures and seven improvements understated the "striking" results the surgery had apparently produced.[14]

Surgery of the Soul

Like Moniz, Walter Freeman was a prominent physician driven by ambition. By 1935, he had an accomplished résumé. Only forty

*Moniz published his last article on lobotomy in 1937. Two years later, he was shot by a disgruntled patient; however, he recovered and continued to practice until 1944, when he retired. He died in 1955, at age eighty-one, six years after he won the Nobel Prize, and in those last years, wrote neurologist António Damásio, he was a man "obviously content with himself."

years old, he was a faculty member at both Georgetown and George Washington University medical schools, the author of a well-received text on neuropathology, and head of the American Medical Association's certification board for neurology and psychiatry, a position that recognized him as one of the leading neurologists in the country. Yet for all that, he could point to no singular achievement. He'd analyzed more than 1,400 brains of the mentally ill at autopsy, intent on uncovering anatomical differences that would explain madness, but had found nothing. This research had proven so barren that Freeman sardonically quipped that whenever he encountered a "normal" brain, he was inclined to make a diagnosis of psychosis. He also was a bit of an odd bird. Brilliant, flamboyant, acerbic, cocky—he wore a goatee and seemed to enjoy prickling the sensibilities of his staid colleagues. He taught his classes with a theatrical flair, mesmerizing his students, in particular with in-class autopsies. Freeman would remove a corpse's skullcap with a saw and then triumphantly remove the brain, holding it up to teach neuroanatomy.[15]

Moniz's surgery had a natural allure for him—it was bold, daring, and certain to ruffle a few professional feathers. It also fit into his own thinking about possible remedies for the mentally ill. Even before Moniz had published his results, he'd suggested, in a paper titled "The Mind and the Body," that brain surgery could find a place in psychiatry's toolbox. Illnesses like encephalitis and syphilis attacked distinct regions in the brain, he'd noted, and those diseases caused alterations in behavior. If a viral agent could change a person's actions, couldn't a neurosurgeon do the same with his knife? "We may be able to influence behavior in a significant manner by destroying localized portions" of the brain, he'd concluded.[16]

Freeman recruited a young neurosurgeon, James Watts, to be his collaborator. Their first patient was, like Moniz's, a sixty-three-year-old woman, A. H. She suffered from severe depression, was suicidal, and obsessed about growing old. Freeman described her as a "master at bitching" who so domineered her husband that he led "a dog's life." Although her family consented to the experiment, she protested that she didn't want any part of it if it would require cutting her hair. Freeman mollified her by assuring her that her precious curls would not be shorn, and on September 14, 1936, he and Watts cut six corings from each of her frontal lobes. The operation went smoothly, and after awaking from anesthesia, A. H. reported that she felt better and that she was no longer sad. She expressed no concern that Freeman had lied to her and that her hair was now gone.[17]

Freeman and Watts wasted no time in announcing their positive results. Before two months had passed, they'd fired off an article to the *Southern Medical Journal*, claiming success. A. H., they said, was now "content to grow old gracefully," was able to manage household chores "as well as she ever did," and enjoyed "the company of her friends who formerly used to exhaust her." Her husband found her "more normal than she had ever been." By the end of the year, Freeman and Watts had operated on sixteen more women and three men. Their published conclusions remained upbeat. Not only did the operation relieve emotional distress, but any intellectual loss was apparently minimal. Memory was described as intact, concentration improved, and judgment and insight undiminished. The patients' ability to enjoy external events had increased. The one negative, Freeman and Watts wrote, was that "every patient probably loses something by this operation, some spontaneity, some sparkle, some flavor of the personality, if it may be so described." But that loss seemed acceptable in patients who "have an otherwise hopeless prognosis," they said.[18]

Freeman proved even better than Moniz at publicizing his and Watts's surgical triumph. Just before he presented the results of their first six surgeries at a meeting of the Southern Medical Society on November 18, 1936, he called a *Washington Star* reporter, Thomas Henry, and gave him an "exclusive." That stirred other reporters into a near frenzy, just as Freeman had hoped. The *New York Times* wrote that their "new operation marked a turning point of procedure in treating mental cases," their work likely to "go down in medical history as another shining example of therapeutic courage." *Time*, *Newsweek*, and other national publications trumpeted their accomplishments as well, and Freeman, enjoying this blush of fame, gradually made ever more startling claims. His new "surgery of the soul," the *New York Times* reported, in a June 7, 1937, article that appeared on its front page, could relieve "tension, apprehension, anxiety, depression, insomnia, suicidal ideas, delusions, hallucinations, crying spells, melancholia, obsessions, panic states, disorientation, psychalgesia (pain of psychic origin), nervous indigestion and hysterical paralysis." The operation, the paper added, "transforms wild animals into gentle creatures in the course of a few hours."[19]

This was astounding stuff. People from around the country sent letters to Freeman and Watts asking about this amazing new operation. If worry, depression, and anxiety could be plucked neatly from the brain, there was no telling what other conditions could be miraculously treated with their amazing leucotomes. Perhaps asthma could be removed from the brain. Or mental retardation? Their very souls apparently could be carved for the better. After the first round of twenty surgeries, Freeman and Watts also altered the operation so that the frontal lobes would be disabled in a more "precise" way.* Instead of drilling into the skull from the top, they cut into the brain from the lateral sides, varying the scope of frontal-lobe damage depending on the patient's diagnosis. For those suffering from emotional disorders, they would make their cuts toward the front of the skull. For those with chronic schizophrenia, they would drill into the frontal lobes farther back. The more posterior the entry point, the larger the portion of the frontal lobes that would, in essence, be disconnected from the rest of the brain.

---

*Although Freeman and Watts told the public they altered their surgery to make it more precise, in truth they were forced to do so because the first twenty operations had, in essence, gone awry. Many of the initial patients had experienced a return of their symptoms and needed repeat operations. One patient had died from a cerebral hemorrhage, another from cardiac arrest not long after the surgery. A third patient, known as Mrs. S. in the literature, who prior to the surgery had worked for thirteen years as a secretary and was married, slid into a profoundly dilapidated state after the operation, from which she never recovered. But the public didn't learn of Mrs. S's miserable fate; she was one of the first six patients to be operated on and was said in the medical journals to be "moderately improved." She was still appearing as a good outcome in the medical literature as late as 1938, two years after her operation, even though by that time she was, in truth, whiling her days away in St. Elizabeth's Hospital in Washington, D.C., "fat, foolish and smiling."[20]

The human mind, it seemed, could be neatly fixed—and even improved—by the surgeon's knife. As Freeman proudly wrote, lobotomy "was a stroke at the fundamental aspect of the personality, that part that was responsible for much of the misery that afflicts man."[21]

The Stamp of Approval

Although the positive results announced by Freeman and Watts created a great stir in psychiatry and in the press, neurosurgeons as a group did not rush to perform the operation. This surgery was clearly a profound one, which gave most physicians great pause. Insulin coma, metrazol, and electroshock may have worked by inflicting trauma on the brain, but there was still much debate over how severe that trauma was or whether it led to permanent damage. With lobotomy, it was clear: This was an operation that permanently destroyed a part of the brain thought to be the center of human intelligence. Did one really dare to do that? With that question hanging in the air, fewer than 300 lobotomies were performed in the United States from 1936 to 1942. But gradually over that period wariness about the operation waned, and it did so for an understandable reason. Nearly all those who tried the operation concluded that it worked wonders.

After Freeman and Watts, the first American neurosurgeon to try lobotomy was James Lyerly, in Jacksonville, Florida. By early 1938, he had performed the surgery on twenty-one patients. The majority he chose for the operation suffered from depression and other emotional disorders, and many had been ill less than a year. He reported spectacular results. Patients who had been painfully worried and anxious had become relaxed and cheerful and were able to laugh once more. They'd gained weight, their "radiant" faces reflecting their new inner happiness. Nor did it appear that such transformation had come at any great cost. In none of the patients, Lyerly wrote, was there any evidence that disconnecting the frontal lobes had affected "the patient's judgment, reasoning, or concentration, or his ability to do arithmetic." They could now "think better and do more work than before." All of the hospitalized patients had either been discharged, or would be soon.[22]

Lyerly presented his results at a meeting of the Florida Medical Association in May 1938, and it convinced his peers that they too needed to start doing the surgery. J. C. Davis, president of the Florida State Board of Medical Examiners, called the outcomes "nothing less than miraculous." Other psychiatrists joined in to praise Lyerly, concluding that the value of such an operation, for patients who otherwise had no hope, "cannot be overrated." All psychiatrists now had an obligation, reasoned P. L. Dodge, to bring this operation "before the rest of the world for the benefit of every patient who suffers from this disease so they might avail themselves of this particular operation." Dodge promised to immediately write the families of his patients and urge them to have their loved ones undergo lobotomy as soon as possible, before they became hopelessly deteriorated.[23]

Other physicians soon reported similar results. Francis Grant, chief of neurosurgery at the University of Pennsylvania, and a close friend of Watts, operated on ten patients at Delaware State Hospital. Seven, he said, had returned home after the surgery. Two of his anecdotal accounts told of remarkable revivals. Prior to the surgery, Sally Gold had been "entirely hopeless." A year later, she was engaged and had invited Grant to attend the wedding. Julia Koppendorf's story was much the same. Before undergoing a lobotomy, she had been so engulfed in depression that her life was "little worth living," Grant said. Twelve months later, her nephew reported that she was now quite normal.[24]

Patients, too, were quoted as singing the praises of the surgery. They were said to write letters of gratitude, detailing their newfound happiness and how their lives had been born anew. Watts received a touching poem from one of his patients.

Gentle, clever your surgeon's hands
God marks for you many golden bands
They cut so sure they serve so well
They save our souls from Eternal Hell
An artist's hands, a musician's too
Give us beauty of color and tune so true
But yours are far the most beautiful to me
They saved my mind and set my spirit free.[25]

Pennsylvania Hospital's Edward Strecker found that the surgery even benefited schizophrenics. Both Moniz and Freeman had determined that it didn't help this group of patients—although they became less emotional, their delusions didn't subside—but Strecker found otherwise. His chronic patients had been miraculously reborn. "Disregard of others," Strecker wrote, "has been replaced by manifestations of thoughtfulness, consideration, and generosity." Artistic and athletic skills were said to be revived. Whereas before the schizophrenic patients had been lost to the world, they now happily thought about the future, eagerly anticipating going on trips, taking cruises, and going to the theater. They scorned the voices that had once tormented them as "silly" and unworthy of heeding.[26]

As had been the case with other published reports, Strecker's anecdotal accounts gripped the imagination. Strecker told of one previously lost soul—in the hospital, she had mutilated herself, wouldn't wear clothes, and had not responded to any other therapies—who had turned into a Good Samaritan hero. While on an outing, she rescued a friend who had been thrown from a horse—applying first aid, stopping a car to get help, accompanying her friend to the hospital, and waiting there until she was out of danger. The disconnection of her frontal lobes had apparently made her both resourceful and compassionate. Another of Strecker's lobotomized patients, a twenty-five-year-old woman, had become the mother of a beautiful baby, was working as a hostess at a resort, and played golf so well that she could compete in highly competitive tournaments. Perhaps most impressive, she had "retained all of her intellectual capacity."

In 1943, Lloyd Ziegler tallied the lobotomy results to date. By that time, there had been 618 lobotomies performed at eighteen different sites in the United States and Canada. Five hundred and eighteen patients were "improved" or "recovered"; 251 were living in the community and working full or part-time. Twelve people had died from the operation. Only *eight* had worsened following the surgery. "We have known for a long time that man may get on with one lung or one kidney, or part of the liver," Ziegler concluded. "Perhaps he may get on, and somewhat differently, with fewer frontal fiber tracts in the brain."[27]

The surgery had passed the test of science. There could no longer be any doubt that the operation greatly benefited the seriously mentally ill.

The Untold Story

Even today, the published study results are stunning to read. The anecdotal reports of lives restored—of hand-wringing, suicidal people leaving hospitals and resuming lives graced by jobs and marriage—are particularly compelling. As they came from physicians with the best credentials, one begins to wonder whether history has been unfairly harsh on lobotomy. We remember it as a mutilating surgery, but perhaps that isn't so. Perhaps it was a worthwhile operation, one that should be revived.

Either that, or there was something missing from the clinical reports.

A fuller view of the effects of lobotomy can be found today, and ironically, it comes from Freeman and Watts. In their 1950 book *Psychosurgery*, they detailed their experiences during more than ten years of performing lobotomies, and as might be expected, they had long-term good news to report. The operation had helped more than 80 percent of the 623 patients they had operated on. Yet it is in this book, which was meant to present lobotomy in a favorable light, that a clear historical picture emerges of just how the surgery transformed the mentally ill. As part of their discussion, Freeman and Watts told families what to expect from patients recovering from lobotomies. Their candid advice, designed to keep families' expectations in check, tells an entirely different story than that depicted in the medical literature.

People who underwent a lobotomy went through various stages of change. In the first weeks following the operation, Freeman and Watts wrote, patients were often incontinent and displayed little interest in stirring from their beds. They would lie in their beds like "wax dummies," so motionless that nurses would have to turn them to keep them from getting bedsores. Relatives would not know what to make of their profound indifference to everything around them, Freeman and Watts said: "[The patient] responds only when they keep after him and then only in grunts; he shows neither distress nor relief nor interest. His mind is a blank . . . we have, then a patient who is completely out of touch with his environment and to whom the passage of time means nothing."[28]

To stir patients, physicians and nurses would need to tickle them, pound on their chests, or grab them by the neck and "playfully throttle" them. When finally prodded to move, patients could be expected to behave in unusual ways. One well-bred lady defecated into a wastebasket, thinking it was a toilet. Patients would "vomit into their soup plates and start eating out of the plate again before the nurse [could] take it away." They would also lose any sense of shame. Patients who were stepping out of the shower or were on the toilet would not be the least bit embarrassed when doctors and nurses came into the bathroom.

In this newly lethargic, shameless state, patients who once had been disruptive to the wards now caused fewer problems. Even patients who had been violent before the operation were likely to behave in quieter ways, Freeman and Watts said.

> We vividly recall a Negress of gigantic proportions who for years was confined to a strong room at St. Elizabeths Hospital. When it came time to transfer her to the Medical Surgical Building for operation five attendants were required to restrain her while the nurse gave her the hypodermic. The operation was successful in that there were no further outbreaks . . . from the day after operation (and we demonstrated this repeatedly to the timorous ward personnel) we could playfully grab Oretha by the throat, twist her arm, tickle her in the ribs and slap her behind without eliciting anything more than a wide grin or a hoarse chuckle.[29]

Lobotomy was to be seen as a "surgically induced childhood." As patients began to stir, they would be given coloring books and crayons. Families were advised to bring them dolls or teddy bears to help keep their simple minds occupied. At times, however, patients recovering from the surgery might stir from their lethargy into overly restless behavior. In that case, Freeman and Watts advised stunning them with electroshock, even as early as a week after the brain surgery. "A few electric shocks may alter the behavior in a gratifying manner . . . When employed, it should be rather vigorous—two to four grand mal seizures a day for the first two days, depending upon the result."[30]

About 25 percent of their patients never progressed beyond this initial stage of recovery and had to remain institutionalized. Some became disruptive again and underwent a second and even a third surgery; each time Freeman and Watts would disconnect a larger section of their frontal lobes. As long as these patients reached a state where they remained quiet and no longer disturbed the wards as they once had, Freeman and Watts would judge them to have had "good" outcomes.

However, the majority of their patients were able to leave the hospital. In the clinical trials, this was seen as conclusive evidence of a positive outcome. What the medical journals failed to detail, though, was the patients' behavior once they returned home. A lobotomized patient was likely to sorely try a family's patience.

The patient's extreme lethargy and lack of initiative were likely to remain present, particularly during the first months. Families would need to pull their loved ones from their beds, as otherwise they might never rise. Freeman and Watts noted that even a full bladder might not rouse the patient:

> It is especially necessary for somebody to pull him out of bed since he won't go to the toilet, and only alertness on the part of those who care for him will prevent a lot of linen going unnecessarily to the laundry. Once the patient has been guided faithfully to the toilet, he may take an hour to complete his business. Then he has to be pulled up off the seat. "I'm doing it," he says. "Just a little while, I'm nearly finished." Usually he finishes in a very little while, but the passage of time means nothing to him and he stays on, not thinking, merely inert. If other members of the family are waiting for the use of the bathroom, this type of behavior can be exasperating.[31]

Families could expect that getting their loved ones dressed, undressed, and bathed would be a chore. They would spend hours in the tub, not washing but, "like little children," spending their time "squirting water around." As they lacked any sense of shame, they sometimes would "present themselves to acquaintances and even strangers inadequately clad." They would likely put on weight, some women getting so fat they would "burst the seams of their dresses and not take the trouble to sew them up." At the table, many would focus single-mindedly on eating, at times grabbing food from the plates of others. This type of behavior, Freeman and Watts cautioned, should be "discouraged from the start."[32]

But efforts to get them to improve their manners were likely to prove futile. "No amount of pleading, reasoning, tears or anger" would do any good. Nor would criticism. Hurl the most insulting epithets at them, Freeman and Watts said, and they would just smile. In fact, "the more insulted they are, the better the patients seem to enjoy it." Even physical abuse might not bother them.

> Patients who have undergone prefrontal lobotomy can stand an enormous amount of maternal overprotection, paternal rejection, sibling rivalry, physical discomfort, strained family situations and loss of loved ones. These happenings in the family constellation make no deep emotional impression upon them . . . occasionally they will cry in response to an external stimulus like the sad part of a movie or a radio act. For themselves and their own sometimes pitiable states, however, they do not mourn. Some patients have taken serious beatings—financial, occupational, even physical—and have come up smiling.[33]

About 25 percent of discharged patients, Freeman and Watts wrote, could be "considered as adjusting at the level of a domestic invalid or household pet." This was not to be seen as a bad outcome, however. These patients, relieved of their mental worries, could now devote their "talents to gossiping with the neighbors or just looking out the window."

> We are quite happy about these folks, and although the families may have their trials and tribulations because of indolence and lack of cooperation, nevertheless when it comes right down to the question of such domestic invalidism as against the type of raving maniac that was operated on, the results could hardly be called anything but good.[34]

Even if the patient had been employed a short time before the surgery, Freeman and Watts still considered the operation to have produced a "fair" result if the patient "becomes a drone, living at home in idleness." They did express regret, however, that some of their patients in this "category of household drone" had been "highly intelligent, gifted, ambitious, and energetic people" who had been operated on a short time after they had fallen ill and, prior to surgery, "had considerable prospects of returning to an active, useful, existence."[35]

Some lobotomized patients did progress beyond this "household pet" level. They were able to become employed again and resume some measure of social life. These were the best outcomes, those who, in the medical literature, were reported to have been miraculously transformed. But, Freeman and Watts cautioned, families shouldn't expect them to do particularly well in their jobs. The only positions that lobotomized patients could hope to take were simple ones that required a "minimum of punctuality, industry, accuracy, compliance, and adaptability." Even a task like keeping house would likely prove too difficult because it required juggling multiple tasks and planning ahead. And while their amiable dispositions might help them land work, they would regularly be fired because "the employer expects a certain amount of production."

Sex was another waterloo. The lobotomized male, Freeman and Watts explained, might begin to paw his wife "at inconvenient times and under circumstances when she may be embarrassed and sometimes it develops into a ticklish situation." His lovemaking was also "apt to be at a somewhat immature level in that the patient seeks sexual gratification without particularly thinking out a plan of procedure." It was up to the woman to learn to enjoy such deficiencies:

> Refusal [of sex] . . . has led to one savage beating that we know of and to several separations. Physical self-defense is probably the best tactic for the woman. Her husband may have regressed to the caveman level, and she owes it to him to be responsive at the cavewoman level. It may not be agreeable at first, but she will soon find it exhilarating if unconventional.[36]

Even at the highest stage of recovery, lobotomized patients could not be expected to provide advice of any merit. Those who had been artists or musicians before becoming ill would never regain much interest in such pursuits. They might play the piano for a while in a mechanical way, but the "emotional exhilaration" that comes from playing would be absent, and eventually they would stop playing altogether. Those who had inventive imaginations before surgery would become "dull and uninspired." People who "previous to operation had been absorbed in their studies of philosophy, psychology, world affairs, medieval history, and so on, find that their preference turns to action stories, murder mysteries, the sports pages and the comics." Nor would they, in their lobotomized state, experience spiritual yearnings, any desire to know God.[37]

Freeman and Watts saw this diminishment as a necessary and even good thing for the mentally ill. Many of their patients had become sick precisely because their minds had been too inventive. Patients who once could find "meaning in the verse of obscure poets" or could imagine what history "would have been like if the early Norsemen had intermarried with the Indians and then descended upon the colonists before they had time to become established" could now live undisturbed by such elaborate mental machinations. Such high-flying imagination, Freeman and Watts wrote, becomes "so entrancing that the individual loses sight of the humdrum pattern of getting an education or earning a living," and if "creative artistry has to be sacrificed in the process, it is perhaps just as well to have a taxpayer in the lower brackets as the result." The person who had once painted pictures, written poetry, or composed music was now "no longer ashamed to fetch and carry, to wait on tables or make beds or empty cans." Their best-outcome patients could be described "as good solid cake but no icing."[38]

Such were Freeman and Watts's description of the behavior of lobotomized patients. Most telling of all, in their book they also reflected on what their patients' behavior revealed about frontal-lobe function. They had now observed hundreds of Phineas Gages. The frontal lobes, they concluded, are the "highest endowment of mankind." It is this area of the brain that gives us consciousness of the self, that allows us to experience ourselves and to project ourselves into the past, present, and future. This is the brain center that allows us to care deeply about who we are and our fate. This is the brain region that stirs creative impulses, ambition, a capacity for love, and spiritual yearnings. The Greeks had been right, Broca had been right, and so had Ferrier and Bianchi. The frontal lobes were what made us uniquely human.

And that's what needed to be taken from the mentally ill.

This mental activity, Freeman and Watts explained, was the source of their suffering. Disconnecting the frontal lobes freed the mentally ill from "disagreeable self-consciousness." It liberated them from "all sense of personal responsibility and of anxious self-questioning as to the ethical rightness of their conduct." The lobotomized person, unable to form a mental picture of the "self," would no long worry about past or future:

> He is freed from anxiety and from feelings of inferiority; he loses interest in himself, both as to his body and as to his relation with his environment, no longer caring whether his heart beats or his stomach churns, or whether his remarks embarrass his associates. His interests turn outward, and obsessive thinking is abolished . . . there is something childlike in the cheerful and unselfconscious behavior of the operated patient.[39]

This was the change described by Freeman and Watts in their first published reports as the loss of a certain "spark" in personality. Lobotomy was not surgery *of* the soul. This was surgery that *removed* the soul. As one critic said, lobotomy was a "partial euthanasia."[40] But the trial results published in the medical journals never captured this sense of profound loss. The journal articles conveyed a different reality, telling in general of an operation that could transform hopelessly lost patients on back wards into happy people, some of whom were working and leading fulfilling social lives.

The question that arises today is what drove the creation of that different reality. Why did those who performed this surgery in the late 1930s and early 1940s see their patients' outcomes through such a rosy lens? For that is clearly what they saw. They perceived this surgery as one that could offer great benefits to the mentally ill.

The Influence of Money

In many ways, the success of lobotomy was foretold before Moniz took up his knife. Ever since the turn of the century, of course, psychiatry had been seeking to transform itself into an academic medical discipline, and that meant it had set its sights on developing modern, science-based treatments. Lobotomy fit this bill perfectly. Brain surgery carried with it the luster of being technologically advanced, born from a keen understanding of how the brain worked. Equally important, the Rockefeller Foundation was providing research funds to produce just this type of success. In the 1920s, the Rockefeller Foundation had identified psychiatry as the medical specialty most in need of reform and had begun providing funding—to the tune of $16 million over the course of twenty years—to achieve this change. Rockefeller money financed new departments of psychiatry at several medical schools. It paid for the creation of research laboratories at the schools as well. Various academic psychiatrists were given money to help introduce new clinical treatments. And the hope, and *expectation,* was that all of these efforts would come together in a classic fashion: Basic research would lead to a better understanding of the biology of the brain, and that knowledge would lead to new treatments. Once the Rockefeller monies started flowing, the clock started ticking—the vision was clear, and Rockefeller-funded scientists could be expected to help achieve it.[41]

One of the Rockefeller-funded scientists was John Fulton. He was chairman of the physiology department at Yale University and directed the laboratory where Carlyle Jacobsen conducted his chimp experiments. Jacobsen had designed his studies to probe frontal-lobe function and to identify *deficits* associated with injury to this region of the brain. He was not investigating whether the frontal lobes might provide a remedy for emotional disorders in humans. However, he had made a casual observation that one of the chimps, Becky, had become calmer after the surgery, and once Moniz reported on his new operation, Fulton spun this observation for his benefit. He told the editor of the *New England Journal of Medicine,* Boston neurologist Henry Viets, that the surgery was "well conceived." Why? Because, Fulton explained, it had been based on animal experiments in his lab that had shown that removing the frontal lobes *prevented* neurosis. This led the journal to editorialize, in 1936, that lobotomy was "based on sound physiological observations" and was a "rational procedure."[42] This same story appeared in a 1938 textbook, and soon it had become an accepted "fact" that Moniz had tried lobotomy only after the chimp experiments had proven that it was likely to work. Fulton even came to believe that story himself, proudly writing in his diary that "the operation had its origin on our lab."[43] By seeing the chimp experiments in this way, Fulton was both grabbing a share of the lobotomy glory for himself and making the point that the Rockefeller money coming to his lab was being well spent.

Another Rockefeller recipient was Edward Strecker, at the University of Pennsylvania. He'd received funds to bring advanced medical treatments into the crowded mental hospitals. Such hospitals were filled with chronic schizophrenics. Those patients were precisely the type that both Moniz and Freeman had found did not benefit from lobotomy, which seemingly would have discouraged Strecker from trying it on them. But he did it anyway, because that is what Rockefeller money expected him to do. And when he concluded that Moniz and Freeman were mistaken, that prefrontal lobotomy benefited this group as well, he—like Fulton—was fulfilling his Rockefeller mandate. Similarly, Washington University in St. Louis, Missouri, had received Rockefeller funding to create a strong program in neurosurgery. After Freeman began reporting positive results with prefrontal lobotomy, the school hired Carlyle Jacobsen as its medical psychologist. He was expected to help Washington University neurosurgeons develop better surgical techniques for lobotomy, a refinement that would minimize the deficits produced by the operation. And like Fulton and Strecker, the Washington University physicians—after fiddling with the surgical methods for the operation—were soon reporting results that indicated the Rockefeller funds were being well spent. From 1941 to 1944, they operated on 101 chronic schizophrenics said to have no hope of recovery and announced that with their improved surgical techniques, fourteen of the patients had been essentially cured, thirty had been able to leave the hospital, and none had become worse. They had developed a method for using lobotomy to help even the most dilapidated schizophrenics.

In short, all of these scientists declared results that worked for *them.* Their announced success ensured that the Rockefeller funds would keep flowing. And collectively, they were each pitching in to tell a story—of basic research producing a breakthrough medical treatment—that signaled psychiatry's arrival as a modern, science-based discipline.

The influence of money can be seen in other ways as well. Neurosurgeons had been waiting for some time for an operation like lobotomy to come along. In the 1930s, they had to scramble for patients. They operated primarily on brain tumors, which were not common enough to provide most neurosurgeons with a prosperous practice. When Watts first set up his practice in Washington, D.C., he told Fulton that he expected it would take years to make the practice profitable. Lobotomy offered neurosurgeons a whole new group of patients to operate on, and it wouldn't be difficult finding them—the state hospitals were filled with hundreds of thousands of people. When Watts presented his initial lobotomy results to the Harvey Cushing Society, which neurosurgeons formed in 1932 to promote their interests, the members responded that "these procedures should be tried."[44] They could hope to earn fees ranging from several hundred dollars to $1,500 for performing a lobotomy, attractive sums to surgeons whose annual salaries at that time might not exceed $5,000. As Harvard Medical School's Stanley Cobb later said: Frontal lobotomy was "returning great dividends to the physiologists. But how great the return is to the patient is still to be evaluated."[45]

State governments also had financial reasons for embracing lobotomy. With more than 400,000 people in public mental hospitals, any therapy that would make it possible to send patients home would be welcomed for the monetary savings it produced. In 1941, Mesroop Tarumianz, superintendent at Delaware State Hospital, calculated this fiscal benefit in detail. He told his peers at an AMA meeting that 180 of the hospital's 1,250 patients would be good candidates for lobotomy; it would cost the state $45,000 to have them operated on. Ten percent could be expected to die as a result of the operation (mostly from cerebral hemorrhages); of the remaining 162 survivors, eighty-one could be expected to improve to the point they could be discharged. All told, the state would be relieved of the care of ninety-nine patients (eighteen deaths and eighty-one discharges), which would produce a savings of $351,000 over a period of ten years. "These figures being for the small state of Delaware, you can visualize what this could mean in larger states and in the country as a whole," Tarumianz told the AMA.[46]

All of these factors fed into each other and encouraged physicians and society alike to see lobotomy in a positive light. There was money to be earned, money to be saved, and professional advancement to be had. But of course that was not the story that psychiatry could tell to itself or to society—everyone would still need to believe that the operation benefited the mentally ill. Those evaluating outcomes would have to find that the patients were better off. They did so for a very simple reason: They framed the question of efficacy, in their own minds, in a way that made it virtually impossible for the surgery to fail.

As various physicians tried the surgery, they routinely described their patients as having no hope of getting well again without the operation. For instance, Francis Grant wrote in his first lobotomy report that agitated depression renders "the life of the victim little worth living" and that without radical intervention, many "can expect no relief from their misery until death intervenes."[47] Wisconsin neurosurgeon David Cleveland said that all fifteen of his first lobotomy patients were "equally hopeless," even though six of the fifteen were under thirty years old, and one was a sixteen-year-old boy, newly ill, whose primary symptoms were "malignant-looking withdrawal" and "silliness."[48] Watts, meanwhile, once answered critics by describing patients operated on as having descended to the level of animals: "They are often naked, refusing to wear clothes, urinate and defecate in the corner. . . . Food is poked through a crack in the door like feeding an animal in a cage."[49]

That perception of the hospitalized mentally ill was accurate in one regard: It did fit prevailing societal views, arising from eugenic beliefs, about the "worth" of the mentally ill. They didn't have any intrinsic value as they were. Nor did people with such bad "germ plasm" have a natural capacity for recovery. And given that starting point for assessing outcomes, any change in behavior that resulted in the patients' becoming more manageable (or less of a bother), could be judged as an improvement. What could be worse than hopeless? At Winnebago State Hospital in Wisconsin, physicians used an outcomes scale that ranged from no change to slight improvement to being able to go home. They didn't even allow for the possibility that patients might become worse. Lyerly used a similar scale: Patients could be seen as "greatly improved, moderately improved, slightly improved and temporarily improved."[50] Their outcome measurements explain why Ziegler, when tallying up the cumulative outcomes for lobotomy patients in 1943, found that 84 percent of the 618 patients had improved, and only 1 percent had "deteriorated." Eugenic conceptions of the mentally ill had provided a baseline for perceiving frontal lobotomy as a rousing success.

A Minor Surgery

Stories of medical success have a way of spinning out of control, and so it was with lobotomy. The results announced by Strecker, Grant, and others led, in the early 1940s, to a new round of feature stories in newspapers and magazines, and the writers and editors didn't spare the hyperbole. "Surgeon's Knife Restores Sanity to Nerve Victims," screamed one headline. "No Worse Than Removing Tooth," said another. "Wizardry of Surgery Restores Sanity to Fifty Raving Maniacs," said a third.[51] The *Saturday Evening Post* compared lobotomy surgeons to master watchmakers, writing that they drilled holes into the brain "at just the right marks, inserting tools very carefully to avoid touching little wheels that might be injured . . . they know the 'works' within the skull."[52] And with the press outdoing itself in this way, the use of lobotomy exploded. Prior to the end of World War II, prefrontal lobotomy had been performed on fewer than 1,000 people in the United States. But over the next decade, more than 20,000 underwent the operation, which also came to be seen as appropriate for an ever-widening circle of patients. Some—mostly women—voluntarily sought it out as a cure for simple depression. College graduates suffering from neurosis or early onset of psychosis were said to be particularly good candidates for the surgery. Freeman and a handful of others tried it as a way to cure troubled children. Most of all, however, it became regularly employed at state mental hospitals.

Freeman acted as the pied piper for this expansion. Not only did he ceaselessly promote its merits, he developed a simplified operating technique—transorbital lobotomy—that made the surgery quicker to perform. Instead of drilling holes in the sides of the patient's head, Freeman attacked the frontal lobes through the eye sockets. He would use an ice pick to poke a hole in the bony orbit above each eye and then insert it seven centimeters deep into the brain. At that point, he would move behind the patient's head and pull up on the ice pick to destroy the frontal-lobe nerve fibers.[53] With this new method, Freeman reasoned it wasn't necessary to sterilize the operating field and waste time with that "germ crap." The use of anesthesia could also be eliminated. Instead, he would knock patients out with electroshock before hammering the ice pick through their eye sockets. This saved time and added a therapeutic element, he believed. The electroshock—three shocks in quick succession—scrambled the "cortical patterns" responsible for psychosis; the surgical destruction of the frontal-lobe tissue then prevented "the patterns from reforming," he said.

Freeman performed his first transorbital lobotomy in 1946. He could do the procedure, which he termed a "minor operation," in less than twenty minutes. With the new approach, intellectual deficits were reduced, he said, and he touted it as a surgery suitable for those who were only mildly ill and not in need of hospitalization. People eager to be relieved of depression or anxiety could undergo the office procedure and leave a few hours later. Freeman's principal advice to families was to bring sunglasses—they would be needed to cover up the patient's black eyes. Other than that, Freeman suggested, patients would likely recover quickly and probably wouldn't even remember having been operated on.

Many families traveled from distant cities to bring their loved ones to Freeman for the quick-fix surgery. The patient's own wishes regarding the operation weren't seen as important; rather, it was the family's interests that were paramount. In fact, Freeman saw resistance in patients—whether they were hospitalized or not—as evidence they were good candidates for lobotomy.

> Some patients come to lobotomy after a long series of exasperating treatments . . . They are still desperate, and will go to any length to get rid of their distress. Other patients can't be dragged into the hospital and have to be held down on a bed in a hotel room until sufficient shock treatment can be given to render them manageable.

> We like both of these types. It is the fishy-handed, droopy-faced individual who grunts an uh-huh and goes along with the family when they take him to the hospital that causes us to shake our heads and wonder just how far we will get.[54]

Soon Freeman was taking his new technique on the road, intent on introducing it to state mental hospitals across the country. Traveling in his station wagon, he spent his summers traveling from asylum to asylum, equipped with a pocket set of ice picks for doing surgery after surgery. In any one day, he might operate on a dozen or more patients, screening records when he arrived and then quickly choosing those he deemed suitable. Practiced as he was by then, he could do the surgery in less than ten minutes and would charge the asylums as little as $25 for each one. To quicken the process, he would drive picks into both eyes at once, rather than one at a time, as he could then step behind the patient and pull on both ice picks to simultaneously destroy tissue in both frontal lobes, thereby shaving a few minutes off the operating time. He would perform so many surgeries in one day that his hands would become sore and his forearms would grow weary.

As part of his routine, Freeman would often train the hospital psychiatrist or psychiatric resident in the procedure. Transorbital lobotomy was so simple, he believed, that even someone with no prior training in surgery could be taught how to do it in a single afternoon. At Millidgeville State Hospital in Georgia, Dr. Lewis Hatcher described his understanding of the technique: "I take a sort of medical icepick, hold it like this, bop it through the bones just above the eyeball, push it up into the brain, swiggle it around, cut the brain fibers like this, and that's it. The patient doesn't feel a thing."[55]

Other physicians who adopted transorbital lobotomy echoed Freeman's argument that it was a minor operation. After conducting more than 100 transorbital procedures at Philadelphia Psychiatric Hospital, Matthew Moore determined that not only could a psychiatrist easily do the operation, but he didn't even need any elaborate equipment or facilities. "It can be stated categorically that if this procedure is ineffectual in helping the patient it will do no harm; the patient may not be improved, but he will not be made worse."[56]

Once lobotomy became commonplace in state asylums, it quickly became used as a treatment for disruptive patients who couldn't be quieted by electroshock. The use of lobotomy at Stockton State Hospital in California, which began in 1947, exemplified this pattern.[57] The first patient lobotomized there was a thirty-three-year-old woman who had undergone 450 electroshock treatments during her first six years at the hospital but still misbehaved. She swore regularly and had poor hygiene. After lobotomy, though, she turned "childlike, naïve, and quite friendly," her new behavior much more pleasing to the staff.

Over the course of the next seven years, 232 patients were lobotomized at Stockton Hospital. California law required that the hospital obtain consent from the patient's family, which was told that the surgery was a "delicate brain operation" and "the most advanced type of treatment that is now available." However, in their chart records, the Stockton doctors privately expressed their real reason for recommending lobotomy: This was an operation that could turn "resistive, destructive" patients into "passive" ones. In 1949, the California Department of Mental Hygiene approvingly noted that lobotomy had been used by Stockton and other state hospitals "chiefly to pacify noisy, assaultive, and uncooperative patients."[58]

The last lobotomy at Stockton Hospital was performed in 1954. Joel Braslow, in his book *Mental Ills and Bodily Cures,* has tallied up the cumulative results: Twelve percent of the patients died from the surgery, mostly because of bleeding in the brain. Many were disabled by seizures, incontinence, and lasting disorientation. By 1960, only 23 percent of the lobotomized patients had been able to leave the hospital, and nobody wanted to provide care for those left on the wards. During the next two decades, as part of the deinstitutionalization process, most were finally discharged to nursing homes. The hospital, putting one last positive spin on the lobotomy era, typically stamped their records with such optimistic conclusions as "improved" and "treatment concluded."[59]

More than 60 percent of all people lobotomized in the United States were patients at state mental hospitals. But like any "successful" procedure, it was eventually tried on children.

In 1950, Freeman and Watts reported that they had operated on eleven troubled youths, including one only four years old.

"The aim has been to smash the world of fantasy in which these children are becoming more and more submerged," they explained. "It is easier to smash the world of fantasy, to cut down upon the emotional interest that the child pays to his inner experiences, than it is to redirect his behavior into socially acceptable channels."[60] Although two of the eleven died, three had to be institutionalized, and three others were described as "antagonistic," "irresponsible," and exhibiting "profound inertia," Freeman and Watts concluded that this first trial in children had produced "modest results," and Freeman continued to occasionally perform such operations throughout the 1950s.

A Eugenic Solution

Medical therapeutics for the mentally ill, and how they are used, invariably reflect underlying societal values. In the 1700s, European societies conceived of the mentally ill as beings that, without their reason, had descended to the level of animals, and they developed harsh therapeutics to tame and subdue them. In the early 1800s, the Quakers in York, England, viewed the mentally ill as brethren, as fellow human beings worthy of their empathy, and fashioned a therapeutic that emphasized kindness and the comforts of a good home. In the first half of the twentieth century, America conceived of the mentally ill as hereditary defectives, without the rights of "normal" citizens. That set the stage for therapeutics that were designed to alter who the mentally ill were, with such remedies to be applied even over their protests.

Insulin coma, metrazol, forced electroshock, and lobotomy all fit this model. Lobotomy simply brought brain-damaging therapeutics—a phrase coined by Freeman—to its logical conclusion. This operation, as physician Leo Alexander pointed out in 1940, was a more precise way to damage the brain:

> There is agreement that the clinical improvement following metrazol or insulin therapy is essentially due to destruction of brain tissue, and that the clinical improvement caused by metrazol or insulin treatment has essentially the same rationale as frontal lobotomy. There can be no doubt, from the scientific point of view,

> that a method in which one knows what parts of the brain are destroyed is preferable to one in which destruction is unpredictable, at random, and more or less left to chance.[61]

In Germany, eugenic attitudes toward the mentally ill led to a euthanasia program. Nazi physicians perceived it as a merciful "medical treatment," and the Nazi government carried it out under the banner of the law, with judges deciding which mentally ill people needed to be "relieved" of the burden of living. In the United States, eugenics led to a different end, but clearly one consistent with eugenic beliefs. It led to a quartet of therapeutics, applied regularly without the patient's consent, that filled the mentally ill with terror, broke their bones, robbed them of their memories, and, in the manner of a partial euthanasia, "relieved" them of the very part of the mind that makes us human. The path to lobotomy, it becomes clear, began not with Moniz but with Charles Davenport and his scorn for the "unfit." Franz Kallman's description of the mentally ill as individuals who were not "biologically satisfactory," the American Eugenics Society's catechism that disparaged the mentally ill as "cancers in the body politic," and the U.S. Supreme Court's 1927 decision authorizing compulsory sterilization of the mentally ill were all stops on the path as well. Metrazol, forced electroshock, and lobotomy were medical solutions consistent with a eugenic conception of the mentally ill.

However, American society has never perceived those treatments in this light. Certainly it did not in the immediate years after World War II. Doctors in Germany, shamed over the revelations at the Nuremberg Doctors Trial, viewed lobotomy with much wariness, seeing it as reminiscent of euthanasia. Freeman's transorbital lobotomy particularly appalled them. But the view was quite different in the United States. The United States was in a triumphant mood, newly confident of its ways, and psychiatry saw in this surgery evidence of its own triumph and arrival as a modern discipline. In 1948, the *American Journal of Psychiatry* proudly commented that "every step of [the pioneers'] progress in this rapidly growing field is marked by a deep sense of primary obligation to the patient, and a profound respect for the human brain."[62] *Mental Hygiene News* adopted a darkened landscape pierced by the light of lobotomy's torch as a symbol for its masthead—lobotomy was the beacon that had so transformed psychiatry. The *New England Journal of Medicine* editorialized that "a new psychiatry may be said to have been born in 1935, when Moniz took his first bold step in the field of psychosurgery."[63] And when Moniz was awarded the 1949 Nobel Prize in medicine and physiology, the *New York Times* hailed the "explorers of the brain" who had invented this "sensational operation."

> Hypochondriacs no longer thought they were going to die, would-be suicides found life acceptable, sufferers from persecution complexes forgot the machinations of imaginary conspirators . . . surgeons now think no more of operating on the brain than they do of removing an appendix . . . it is just a big organ with very difficult and complicated functions to perform and no more sacred than the liver.[64]

The tale America had been telling itself had wound its way to a wholly satisfying conclusion: Lobotomy was the fruit of both good science and a humanitarian empathy for the mentally ill.

PART THREE

BACK TO BEDLAM

(1950–1990s)

6

MODERN-DAY ALCHEMY

The drug produced an effect similar to frontal lobotomy.
—N. William Winkelman Jr. (1954)[1]

THE MODERN ERA of medical treatments for schizophrenia is always traced back to a specific date: May 1954. That month, Smith, Kline & French introduced chlorpromazine into the U.S. market, selling it as Thorazine. This drug was the first "antipsychotic" medication to be developed, and it is typically remembered today as dramatically different in kind from lobotomy and the other brain-disabling therapies that preceded it. In his 1997 book *A History of Psychiatry,* Edward Shorter neatly summed up this belief: "Chlorpromazine initiated a revolution in psychiatry, comparable to the introduction of penicillin in general medicine." With this drug, Shorter added, schizophrenia patients "could lead relatively normal lives and not be confined to institutions."[2]

But that was not at all how chlorpromazine was viewed in 1954. It was seen at that time as a pill that hindered brain function, much in the same manner that lobotomy did. It took a decade of modern-day alchemy to turn it into the "antipsychotic" medication we recall today.

First Impressions

Although eugenics had become a thoroughly shamed science by the 1950s, intimately associated with the horrors of Nazism, the therapeutics it had spawned didn't suddenly disappear. Approximately 10,000 mental patients in the United States were lobotomized in 1950 and 1951, which was nearly as many as had been operated on during all of the 1940s. Electroshock remained a mainstay treatment in state hospitals, and it was often used to deliberately reduce patients to confused states. In 1951, for instance, psychiatrists at Worcester State Hospital in Massachusetts reported that they had successfully used repetitive electroshock to "regress" fifty-two schizophrenics to the point where they were incontinent, unable to feed or dress themselves, and mute. D. Ewen Cameron, who was named president of the American Psychiatric Association in 1952, also utilized electroshock in this way, shocking his patients up to twelve times daily, which, he wrote, produced a disruption in memory "so massive and pervasive that it cannot well be described." Patients so treated, he said, were unable even to "conceptualize" where they were. Nor did eugenic sterilizations cease. Approximately 4,000 mentally ill patients were sterilized in the 1950s, which was about the same number as in the 1920s, when eugenic attitudes toward the mentally ill were reaching a feverish pitch. This was the therapeutic milieu that was still in place—the value system, as it were—when chlorpromazine made its debut in the state mental hospitals.[3]

Chlorpromazine, which was synthesized in 1950 by Rhône-Poulenc, a French pharmaceutical firm, belonged to a class of compounds, known as phenothiazines, that were developed in the late 1800s for use as synthetic dyes. In the 1930s, the U.S. Department of Agriculture employed phenothiazine compounds for use as an insecticide and to kill swine parasites. Then, in the 1940s, phenothiazines were found to sharply limit locomotor activity in mammals, but without putting them to sleep. Rats that had learned to climb ropes in order to avoid painful electric shocks could no longer perform this escape task when administered phenothiazines.

This effect inspired investigations by French researchers into whether phenothiazines could be used during surgery to enhance the effects of barbiturates and other anesthetics—perhaps phenothiazines could numb the central nervous system in a novel way. Rhône-Poulenc experimented with various phenothiazine derivatives before selecting chlorpromazine as one that might best achieve this numbing effect.

In 1951, French naval surgeon Henri Laborit tested chlorpromazine on surgical patients and found that it worked so well operations could be performed with almost no anesthesia. He also observed that it put patients into an odd "twilight" state. They would become emotionally detached and disinterested in anything going on around them, yet able to answer questions. One of Laborit's colleagues likened this effect to a "veritable medicinal lobotomy," an observation that suggested it might have use in psychiatry.[4]

A year later, French psychiatrists Jean Delay and Pierre Deniker announced that they had used it to calm manic patients at St. Anne's Hospital in Paris. It was just as Laborit had said. Chlorpromazine induced in patients a profound indifference. They felt separated from the world "as if by an invisible wall."

> Seated or lying down, the patient is motionless on his bed, often pale and with lowered eyelids. He remains silent most of the time. If questioned, he responds after a delay, slowly, in an indifferent monotone, expressing himself with few words and quickly becoming mute. Without exception, the response is generally valid and pertinent, showing that the patient is capable of attention and of reflection. But he rarely takes the initiative of asking a question; he does not express his preoccupations, desires, or preference. He is usually conscious of the amelioration brought on by the treatment, but he does not express euphoria. The apparent indifference or the delay in response to external stimuli, the emotional and affective neutrality, the decrease in both initiative and preoccupation without alteration in conscious awareness or in intellectual faculties constitute the psychic syndrome due to the treatment.[5]

Delay and Deniker dubbed their new treatment "hibernation therapy." Other European psychiatrists soon found it useful for the same reason. Chlorpromazine, they announced, produced a "vegetative syndrome" in patients. Psychotic patients on chlorpromazine became "completely immobile" and could be "moved about like puppets." British psychiatrist D. Anton-Stephens found that drugged patients "couldn't care less" about anything around them and would lie "quietly in bed, staring ahead"—a bother to no one in this drugged state.[6]

The first psychiatrist in North America to test chlorpromazine was Heinz Lehmann, at Verdun Protestant Hospital in Montreal. Like his European peers, Lehmann speculated that it "may prove to be a pharmacological substitute for lobotomy." Medicated patients became "sluggish," "apathetic," "disinclined to walk," less "alert," and had an empty look—a "vacuity of expression"—on their faces. They spoke in "slow monotones." Many complained that chlorpromazine made them feel "empty" inside, Lehmann noted. "Some patients dislike the treatment and complain of their drowsiness and weakness. Some state that they feel 'washed out,' as after an exhausting illness, a complaint which is indeed in keeping with their appearance."[7]

U.S. psychiatrists initially perceived chlorpromazine's effects this way as well. The drug, wrote Philadelphia psychiatrist N. William Winkelman Jr., transformed patients. Those "who had been severely agitated, anxious and belligerent became immobile, waxlike, quiet, relaxed and emotionally indifferent."[8] Texas psychiatrist Irvin Cohen reported: "Apathy, lack of initiative and loss of interest in surroundings are a common response in patients."[9] In 1955, Deniker and Delay coined the term "neuroleptic" to describe the effects produced by chlorpromazine and other phenothiazines that had been introduced. The word came from the Greek, meaning to "take hold of the nervous system," reflective of how the drugs were perceived to act as chemical restraints.

Very early on, physicians in Europe and the United States realized that chlorpromazine frequently induced Parkinson's disease symptoms—the shuffling gait, the masklike visage, and even the drooling. Swiss psychiatrist Hans Steck announced in 1954 that 37 percent of the 299 mental patients he'd treated with chlorpromazine showed signs of Parkinson's.[10] Lehmann noticed the same thing. In the United States, more than 100 psychiatrists who met in Philadelphia in June 1955 spoke at great length about this side effect. "Our feeling has been that all patients who are on large doses of Thorazine for any length of time show some signs of basal ganglion dysfunction," noted George Brooks, from Vermont State Hospital. "Not perhaps full-blown Parkinsonism, but some loss of associated movements, loss of facial mobility, etc." Hyman Pleasure, a psychiatrist from Pilgrim State Hospital in New York, reported the same findings: "Probably two-thirds of our patients showed some degree of Parkinson-like symptoms." Added Delaware State Hospital psychiatrist Fritz Freyhan: Chlorpromazine can "metamorphose a highly mobile, flighty manic into a static, slow-motion shuffler."[11]

Indeed, Freyhan and others at the 1955 meeting debated whether such symptoms should be deliberately induced. Many observed that the best therapeutic results, in terms of producing an emotional tranquillity in patients, coincided with the appearance of the motor disability. This led some to speculate that Parkinson's was somehow antagonistic to schizophrenia, much in the same way that convulsions had once been thought to chase away the disorder. If so, the proper therapeutic dosage would be one that induced this motor disability, and then perhaps the symptoms of this disease could be controlled with other drugs. Winfred Overholser, superintendent of St. Elizabeth's Hospital in Washington, D.C., closed the Philadelphia symposium with this question for his peers: "Should you push the drug to the stage of bringing about Parkinsonism? Is it a fact that the ratio of improvement or symptomatic recovery is greater in the cases in which Parkinsonism is developed?"[12]

During this initial period, psychiatrists did not perceive chlorpromazine as having any specific antipsychotic properties. "It is important to stress that in no case was the content of the psychosis changed," wrote England's Joel Elkes, in 1954. "The schizophrenic and paraphrenic patients continued to be subject to delusions and hallucinations, though they appeared to be less disturbed by them."[13] Instead, neuroleptics were perceived to "work" by hindering brain function. Chlorpromazine, Lehmann observed, has the "remarkable property of inhibiting lower functional centers of the central nervous system without significantly impairing the function of the cortex."[14] Laborit said that the drug's principal therapeutic effect was the "disconnection of the neurovegetative system."[15] Animal experiments showed that lesions in the caudal hypothalamus produced similar deficiencies in motor skills and initiative. Neuroleptics, researchers concluded, "modified" patients in ways that made their behavior more acceptable to others. They could be used "to attain a neuropharmacologic effect, not to 'cure' a disease."[16]

By 1957, Delay and Deniker had also recognized that neuroleptics produced deficits similar to those caused by encephalitis lethargica. This ailment, which struck 5 million people worldwide during a 1916–1927 epidemic, caused a brain inflammation that left people apathetic, lacking the will to do anything, and with waxlike facial expressions. Physicians described the disease, known colloquially as sleeping sickness, as causing "psychomotor inertia." Chlorpromazine caused eerily similar deficits, only at a much faster pace. Deniker wrote: "It was found that neuroleptics could experimentally reproduce almost all the symptoms of lethargic encephalitis. In fact, it would be possible to cause true encephalitis epidemics with the new drugs."[17]

Although it might seem strange today that a drug described in this manner would be welcomed into the state mental hospitals, at the time such effects were seen as *desirable*. In the early 1950s, insulin coma, electroshock, and frontal lobotomy were all perceived as helpful therapies. The asylum conditions that had led to those earlier brain-disabling therapies being declared effective—did they make patients quieter, easier to manage, and less hostile?—were also still in place. In 1954, hospital administrators were still struggling with horribly inadequate budgets and hopelessly overcrowded facilities. A drug that could reliably tranquilize disruptive patients was bound to be welcomed. Hospital staff—much in the same way they had felt more kindly toward patients reduced to childlike behavior by insulin coma—even felt more empathetic toward their patients once they were stilled by chlorpromazine.

"Chlorpromazine [has] produced a decrease in brutality in mental hospitals which was not achievable by any system of supervision or control of personnel," declared Anthony Sainz of Marcy State Hospital in New York. "Many patients, for example, when they develop a central ganglionic or Parkinsonian syndrome become more 'sick' and thus arouse the sympathies of those taking care of them instead of arousing their anger and hostility. The patients, in consequence, receive better care rather than worse."[18]

Chlorpromazine, then, initially found a place *within* asylum medicine. However, even as it was making its debut in that environment, the United States was in the first stage of rethinking its care of the mentally ill and envisioning a change that would, at least in theory, require a pill different in kind from chlorpromazine. With eugenics now a shamed science, there was no longer the same societal belief that the mentally ill necessarily needed to be segregated, and yet the states were still stuck with the financial consequences of that eugenics legacy. There were more than 500,000 people in public mental institutions, and even though states were still scrimping on expenses, spending less than $2 per day per patient (less than one-seventh the amount spent in general hospitals), their collective expenditures for the mentally ill had reached $500 million annually. They wanted to get out from under that expensive burden, and in the early 1950s, the Council of State Governments, which had been meeting annually to discuss this problem, articulated a vision of reform. "There are many persons in state hospitals who are not now in need of continuing psychiatric hospital care," the council announced. "Out-patient clinics should be extended and other community resources developed to care for persons in need of help, but not of hospitalization."[19]

America had a new agenda on the table, replacing asylum care with community care. But for that agenda to proceed, America would need to believe that a medical treatment was available that would enable the seriously mentally ill to *function* in the community. The neuroleptics that had been embraced in asylum medicine—drugs that reliably made patients lethargic, emotionally disengaged, and retarded in movement—hardly fit that bill. A pill of a different sort would be needed, and so it was, with that fiscal agenda on the table, that neuroleptics, over the course of ten years, underwent a remarkable transformation.

Spinning Dross into Gold

In one manner or another, mad medicine is always shaped by larger forces coursing through a society. The brain-damaging somatic therapies of the 1930s—insulin coma, electroshock, and lobotomy—all appeared in asylum medicine while American society was under the influence of eugenics. Chlorpromazine made its debut as a successor to those therapies, and then its image was transformed in a society newly under the influence of pharmaceutical money.

After World War II, global leadership in drug development began to shift from Germany to the United States, and it did so because the financial opportunities in the United States were so much greater. Drug manufacturers in the United States could get FDA approval for their new medications with relative ease, since at that time they did not have to prove that their drugs were effective, only that they weren't too toxic. They could also charge much higher prices for their drugs in the United States than in other countries because of strong patent-protection laws that limited competition. Finally, they could count on the support of the influential American Medical Association, which, as a result of a new law, had begun cozying up to the pharmaceutical industry.

Prior to 1951, the AMA had acted as a watchdog of the drug industry. In the absence of government regulations requiring pharmaceutical companies to prove that their medications had therapeutic merit, the AMA, for nearly fifty years, had assumed the responsibility of distinguishing good drugs from the bad. It had its own drug-testing laboratory, with drugs deemed worthwhile given the AMA seal of approval. Each year it published a book listing the medications it found useful. Drug companies were not even allowed to advertise in the *Journal of the American Medical Association* unless their products had been found worthy of the AMA seal. At that time, however, patients could obtain most drugs without a doctor's prescription. Drug companies primarily sold their goods directly to the public or through pharmacists. Physicians were not, in essence, drug vendors. But in 1951, Minnesota senator Hubert Humphrey cosponsored a bill, which became the Durham-Humphrey Amendment to the Federal Food, Drug and Cosmetics Act of 1938, that greatly expanded the list of medications that could be obtained only with a doctor's prescription. While the amendment was designed to protect the public by allowing only the safest of drugs to be sold over the counter, it also provided doctors with a much more privileged status within society. The selling of nearly all potent medications now ran directly through them. As a result, drug companies began showering them, and their professional organizations, with their marketing dollars, and that flow of money changed the AMA almost overnight.

In 1950, the AMA received $5 million from member dues and journal subscriptions but only $2.6 million from drug-company advertisements in its journals. A decade later, its revenue from dues and subscriptions was still about the same ($6 million), but the money it received from drug companies had leaped to $10 million—$8 million from journal advertisements and another $2 million from the sale of mailing lists. As this change occurred, the AMA dropped its critical stance toward the industry. It stopped publishing its book on useful drugs, abandoned its seal-of-approval program, and eliminated its requirement that pharmaceutical companies provide proof of their advertising claims. In 1961, the AMA even opposed a proposal by Tennessee senator Estes Kefauver to require drugmakers to prove to the Food and Drug Administration (FDA) that their new drugs were effective. As one frustrated physician told Kefauver, the AMA had become a "sissy" to the industry.[20]

But it wasn't just the AMA that was being corrupted. Starting in 1959, Kefauver directed a two-year investigation by the Senate Subcommittee on Antitrust and Monopoly into drug-industry practices, and his committee documented how the marketing machinery of pharmaceutical firms completely altered what physicians, and the general public, read about new medications. Advertisements in medical journals, the committee found, regularly exaggerated the benefits of new drugs and obscured their risks. The "scientific" articles provided a biased impression as well. Prominent researchers told Kefauver that many medical journals "refused to publish articles criticizing drugs and methods, lest advertising suffer." Pfizer physician Haskell Weinstein confessed that pharmaceutical companies ghostwrote many of the laudatory articles:

> A substantial number of the so-called medical scientific papers that are published on behalf of these drugs are written within the confines of the pharmaceutical houses concerned. Frequently the physician involved merely makes the observations and his data, which sometimes are sketchy and uncritical, are submitted to a medical writer employed by the company. The writer prepares the article

> which is returned to the physician who makes the overt effort to submit it for publication. The article is frequently sent to one of the journals which looks to the pharmaceutical company for advertising and rarely is publication refused. The particular journal is of little interest inasmuch as the primary concern is to have the article published any place in order to make reprints available. There is a rather remarkable attitude prevalent that if a paper is published then its contents become authoritative, even though before publication the same contents may have been considered nonsense.[21]

In its 1961 report, Kefauver's committee also detailed how pharmaceutical companies manipulated the popular press. Magazines were promised advertising revenues if they would publish features mentioning a company's drug in a positive light. Writers could earn extra fees on the side for doing the same, with one scribe telling of a potential payoff of $17,000—far more than a year's salary at the time—for a single magazine article. Writers were also bribed with free dinners, limousine rides, and other perks. Weinstein told Kefauver's committee that, as with the scientific literature, "much of what appears (in the popular press) has in essence been placed by the public relations staffs of the pharmaceutical firms. A steady stream of magazine and newspaper articles are prepared for distribution to the lay press."[22]

In short, in the 1950s, what American physicians and the general public learned about new drugs was molded, in large part, by the pharmaceutical industry's marketing machine. This molding of opinion, of course, played a critical role in the recasting of neuroleptics as safe, *antischizophrenic* drugs for the mentally ill.

Smith, Kline & French obtained the rights to market chlorpromazine in the United States from Rhône-Poulenc in the spring of 1952. At that time, it wasn't a large pharmaceutical house and had annual sales of only $50 million. While it foresaw many possible therapeutic uses for chlorpromazine, it wanted to get the drug on the market as quickly as possible and thus tested it primarily as an antivomiting agent. All told, the company spent just $350,000 developing the drug, administering it to fewer than 150 psychiatric patients for support of its new drug application to the FDA. "Let's get this thing on the market as an anti-emetic," reasoned the company's president, Francis Boyer, behind closed doors, "and we'll worry about the rest of that stuff later."[23]

The FDA approved chlorpromazine on March 26, 1954, and a few days later Smith Kline fired the first shot in its marketing campaign. It produced a national television show, titled "The March of Medicine," and now it was time to craft a story of dutiful science at work. Thorazine, Boyer told the American public, had been rigorously tested:

> It was administered to well over 5,000 animals and proved active and safe for human administration. We then placed the compound in the hands of physicians in our great American medical centers to explore its clinical value and possible limitations. In all, over 2,000 doctors in this country and Canada have used it . . . the development of a new medicine is difficult and costly, but it is a job our industry is privileged to perform.[24]

The television show was the kickoff in an innovative, even brilliant plan for selling the drug. In order to woo state legislatures, which would need to allot funding for use of the drug in mental hospitals, Smith, Kline & French established a fifty-member task force, with each member assigned to a state legislature. The task force organized a "speakers' training bureau" to coach hospital administrators and psychiatrists on what to say to the press and to state officials—a public message of a breakthrough medication needed to be woven. There would be no comments about chemical lobotomies or encephalitis lethargica. Instead, a story of lost lives being wonderfully restored would be told. The company also compiled statistics on how use of the drug would save states money in the long run—staff turnover at asylums would be reduced because handling the patients would be easier, facility maintenance costs would be decreased, and ultimately, at least in theory, many medicated patients could be discharged. This was a win-win story to be created—the patients' lives would be greatly improved and taxpayers would save money.

With the company's training bureau at work in this way, chlorpromazine underwent a step-by-step transformation in the popular press, and in the medical literature as well.

In June 1954, *Time* published its first article on chlorpromazine. At that point, the task force had just set up shop, and so the makeover of chlorpromazine's image was still at an early stage. In an article titled "Wonder Drug of 1954?" *Time* reported:

> After a few doses, says Dr. Charles Wesler Scull of Smith, Kline & French, patients who were formerly violent or withdrawn lie "molded to the bed." When a doctor enters the room, they sit up and talk sense with him, perhaps for the first time in months. There is no thought that chlorpromazine is any cure for mental illness, but it can have great value if it relaxes patients and makes them accessible to treatment. The extremely agitated or anxious types often give up compulsive behavior, a surface symptom of their illness. It is, says Dr. Scull, as though the patients said, "I know there's something disturbing me, but I couldn't care less."[25]

While filled with praise for chlorpromazine, the article still did not describe medicated patients as being "cured" or walking about with great energy. This was still a chemical agent that "molded" patients to the bed and induced emotional indifference. But over the course of the next twelve months, as can be seen in coverage by the *New York Times,* the story being fed to the press changed. Researchers started hinting that chlorpromazine might be curative, a pill that quickly healed the mind and enabled people to go about their daily business in normal fashion.

In 1955, the *New York Times* reported on chlorpromazine at least eleven times. "New Cure Sought for Mentally Ill" ran one headline. "Drug Use Hailed in Mental Cases" said another. The theme repeated over and over was this: Chlorpromazine was "one of the most significant advances in the history of psychiatric therapy." Hospitals using the drug were releasing patients "at a record rate." This was a "miracle" pill that would make it possible for family doctors to treat mental illness in their offices, with "only the most seriously disturbed" needing to be hospitalized. Chlorpromazine brought the disturbed patient "peace of mind" and "freedom from confusion." Virtually nothing was said about the drug's side effects; not one of the eleven articles mentioned that it caused Parkinson's symptoms or lethargy.[26] On June 26, 1955, *New York*

Times medical writer Howard Rusk confidently declared that the neuroleptics had proven their worth:

> Today, there can be little doubt that, in the use of these and other drugs under study, research has developed new tools that promise to revolutionize the treatment of certain mental illnesses. [The drugs] gradually calm patients, who then lose their fear and anxiety and are able to talk about their troubles more objectively. *Patients do not develop the lethargy that follows the use of barbiturates* . . . there is no doubt of the effectiveness of these new drugs in either curing or making hitherto unreachable patients amenable to therapy. (italics added)[27]

Psychiatric researchers also saw an opportunity to use this tale of medical progress to lobby Congress for increased research funds. In May 1955, Nathan Kline, Henry Brill, and Frank Ayd told a Senate budget committee that neuroleptics had given the field new hope. Thanks to the tranquilizers, they said, "patients who were formerly untreatable within a matter of weeks or months become sane, rational human beings." Hospitalization could "be shortened, often avoided altogether." Their lobbying led *U.S. News and World Report* to announce that new "wonder drugs" were "promising to revolutionize the treatment of mental disease."[28] *Time* even suggested that neuroleptics marked a medical advance as profound as the "germ-killing sulfas discovered in the 1930s." Physicians who resisted using them, it added, were "ivory-tower critics" who liked to waste their time wondering whether a patient "withdrew from the world because of unconscious conflict over incestuous urges or stealing from his brother's piggy bank at the age of five."[29]

Not surprisingly, with this storytelling at work, Congress coughed up. Federal spending on mental-health research rose from $10.9 million in 1953 to $100.9 million in 1961—a tenfold increase in eight years. The storytelling also gave state legislators real hope that community care could replace hospital care. At the governors' conference in 1955, the states pledged support "for a full-scale national survey on the status of mental illness and health in the light of new concepts and treatment methods."[30] That same year, Congress passed the Mental Health Study Act, which established the Joint Commission on Mental Illness and Mental Health to devise a plan for remaking the nation's care of the mentally ill.

This image makeover of chlorpromazine in the lay press was being repeated, to some extent, in the medical literature. In the first decade after its approval, more than 10,000 articles in medical journals discussed it. Most were laudatory. And once a new public story began swirling around the drug, many investigators changed their first impressions. William Winkelman's first two published reports illustrate this change.

In 1953, when Smith, Kline & French chose Winkelman to be its lead investigator on its initial tests of chlorpromazine, surgical lobotomy was still seen as a good thing. It was the therapy that chlorpromazine had to measure up to, and when Winkelman reported his initial results, in the *Journal of the American Medical Association* on May 1, 1954, he praised the drugs for being similar in kind. "The drug produced an effect similar to frontal lobotomy," he said approvingly. It made patients "immobile," "waxlike," and "emotionally indifferent."[31] However, three years later, in a study of 1,090 patients published in the *American Journal of Psychiatry*, Winkelman painted a new picture. Motor dysfunction was suddenly nowhere to be found. In this large cohort of patients, followed for up to three years, Winkelman said that he had "not seen a full-blown case of Parkinsonism."[32] Only two of the 1,090 patients even showed faint signs of this disorder, he said. This, of course, was a remarkable change from the talk at the Philadelphia symposium two years earlier, when one physician, Brooks from Vermont, had seen evidence of Parkinsonism in *all* of his patients. But it fit in well with the story being told in the popular press of hopeless patients suddenly being returned to normal, or, as in the case of the *New York Times*, the story of a drug that didn't cause lethargy.

The AMA, meanwhile, also stepped in to ensure that this story of medical progress was not derailed.

There were any number of psychiatrists who were dismayed by the glowing reports of chlorpromazine in the press and medical literature. One, writing in the *Nation*, described it as "vulgarized falsity."[33] Gregory Zilboorg, a prominent New York psychoanalyst, blasted the press, saying that the public was being egregiously misled and that the only real purpose of the drug was to make hospitalized patients easier to handle. "If I hit you over the head and make you bleary eyed," he asked rhetorically, "will you understand me better?"[34] Yet another well-known physician, Lawrence Kolb, who had formerly directed the U.S. Public Health Services' mental-hygiene division, called neuroleptics "physically more harmful than morphine and heroin."[35] Such criticism made for an almost bizarre public confusion. Were neuroleptics wonder drugs or not? Even Kline and Ayd, who'd told their own wonder story to Congress, complained that drugmakers were making false claims in their advertisements and mailings. A House subcommittee decided to investigate, and it was then, with the industry on the hot seat, that the AMA rushed to its defense. Drug companies were acting responsibly with their advertisements, Dr. Lee Bartemeier, chairman of the AMA's committee on mental health, told the House.[36] They were not heaping "extravagant and distorted literature" on the nation's physicians. His testimony defused the matter, and no one put two and two together when, in the following months, the AMA launched *Archives of General Psychiatry,* its pages filled with advertisements for the new miracle drugs.

Smith, Kline & French could certainly afford the marketing expense. In 1958, *Fortune* magazine ranked it second among 500 American industrial corporations in terms of highest "net profit after taxes on invested capital," with its whopping return of 33.1 percent. Its high profit margins reflected the fact that it was charging $3.03 for a bottle of chlorpromazine, six times what Rhône-Poulenc, the inventor of the drug, could charge in France.[37] Some states were now spending approximately 5 percent of their mental-hospital budgets for Thorazine. Indeed, Smith, Kline & French's payoff from its $350,000 investment in chlorpromazine was one for the record books. The company's revenues skyrocketed from $53 million in 1953 to $347 million in 1970, with Thorazine contributing $116 million that year alone.[38]

The Delusion Is Complete

In early 1963, President John Kennedy unveiled his plan for reforming the nation's care of the mentally ill. The state hospitals, relics from a shameful past, would be replaced by a matrix of community care, anchored by neighborhood clinics. At the heart of this vision, the medical advance that made it possible, were the neuroleptics. Two years earlier, Kennedy had received the recommendations of the Joint Commission on Mental Illness and Mental Health, and in that report the drugs had been described as having "delivered the greatest blow for patient freedom, in terms of nonrestraint, since Pinel struck off the chains of the lunatics in the Paris asylum 168 years ago . . . In the surprising, pleasant effects they produce on patient-staff relationships, the drugs might be described as moral treatment in pill form."[39] Kennedy drove home the point for the American people: The new drugs made "it possible for most of the mentally ill to be successfully and quickly treated in their own communities and returned to a useful place in society."[40]

Two critical studies had put the final stamp of science on this belief. The first consisted of a series of reports by Henry Brill and Robert Patton, employees of the New York State Department of Mental Hygiene, assessing whether neuroleptics had led to a decline in the patient census at the state's mental hospitals. Nationwide, the patient census had declined from 558,600 in 1955 to 528,800 in 1961. In New York, the census had dropped from 93,314 in 1955 to 88,764 in 1960—evidence, many argued, that the neuroleptics were helping people get well. However, as Brill and Patton acknowledged, isolating neuroleptics as the specific cause of that slight decline was quite difficult. Hospitalization rates for the mentally ill always reflect social policies—should the mentally ill be quarantined or not?—and by 1954, states were shouting that the patient census needed to drop. New York and many other states, in fact, had begun developing community care initiatives in the early 1950s, funneling the mentally ill into nursing homes, halfway houses, and sheltered workshops. In spite of these confounding factors, Brill and Patton concluded that neuroleptics must have played at least some role in the decline, since the drop in census, however slight, coincided with the introduction of neuroleptics. The fact that the two occurred at the same time was seen as the proof.[41]

Their work became widely cited, and was much discussed by the Joint Commission in its report. But in their research, Brill and

Patton hadn't compared discharge rates for drug-treated versus nontreated patients, a shortcoming that became evident when investigators at California's mental hygiene department did precisely that. In a study of 1,413 first-episode male schizophrenics admitted to California hospitals in 1956 and 1957, they found that "drug-treated patients tend to have longer periods of hospitalization . . . furthermore, the hospitals wherein a higher percentage of first-admission schizophrenic patients are treated with these drugs tend to have somewhat higher retention rates for this group as a whole."[42] In short, the California investigators determined that neuroleptics, rather than speeding people's return to the community, apparently *hindered* recovery. But it was the Brill and Patton research that got all of the public attention. Their conclusions supported the story that the public wanted to hear.

The second study that made Kennedy's plan seem feasible was a multi-site trial of neuroleptics led by the National Institute of Mental Health (NIMH). While the medical journals in the 1950s may have filled up with articles lauding the new drugs, the research behind the articles was recognized as mostly pap: Few convincing placebo-controlled, double-blind studies—a trial design that had come to be recognized as a standard for good drug research—had been conducted. In 1961, the NIMH launched a nine-hospital study, evaluating outcomes in newly admitted patients over a six-week period, to remedy this deficiency. The announced results were stunning. None of the 270 drug-treated schizophrenics became worse, 95 percent improved somewhat, and nearly 50 percent improved so dramatically that they could be classified as either "normal" or only "borderline ill." Indeed, the NIMH-funded investigators concluded that chlorpromazine and two other neuroleptics reduced apathy, improved motor movement, and made patients less indifferent—precisely the *opposite* conclusions drawn by their peers a decade earlier. Side effects, meanwhile, were said to be "mild and infrequent . . . more a matter of patient comfort than of medical safety." Most convincing of all, the NIMH determined that the drugs were indeed curative: "Almost all symptoms and manifestations characteristic of schizophrenic psychoses improved with drug therapy, suggesting that the phenothiazines should be regarded as 'antischizophrenic' in the broad sense. In fact, it is questionable whether the term 'tranquilizer' should be retained."[43]

The transformation of the neuroleptics was now complete. A drug that when first introduced was described as a chemical lobotomy, useful for making patients sluggish and emotionally indifferent, had become a safe and effective medication for schizophrenia. And that clearly is what the psychiatrists who participated in the NIMH trial now honestly saw. Their perceptions had changed in ways that matched societal goals and the story fashioned by drug companies over the past decade. Pharmaceutical ads, the flood of published articles in the scientific literature, the many stories in the popular media of miracle drugs—all had told of drugs that could heal the mentally disturbed. That was the belief that had been crafted, and, in the NIMH trial, the investigators had made observations consistent with it. They saw, in the altered behavior of their medicated patients, the image of their own expectations.

It was also a "reality" that worked for many. The states had wanted to shed the financial burden of their public mental hospitals, and now a scientific rationale was in place for discharging patients into the community. Psychiatry could now pride itself on having become a fully modern discipline, able to offer patients curative pills. Pharmaceutical companies, meanwhile, could count on states to set up programs focused on medicating discharged patients. Rather than serving as a short-term remedy for calming manic patients, neuroleptics were now medications that needed to be taken continuously. Pharmaceutical firms had lifelong customers for their drugs, and a society poised to insist that such drugs be taken. Finally, in this optimistic time of Kennedy's Camelot, American society could believe it was righting yet another social abuse from the past. The mentally ill, so long neglected, would now be welcomed into the community. As Wilbur Cohen, acting secretary of the Department of Health, Education and Welfare, said a few years later, many among the mentally ill "can be put back to work and can be given a rightful place in society, and they are not a drain on either their families or the taxpayer."[44]

Unfortunately, it was a good-news tale that was missing one key voice: that of the mentally ill. There had been little mention of how they felt about these wonder drugs. It was a glaring absence, and, as usual, their perceptions were quite at odds with society's belief that a safe "antischizophrenic" treatment had been found. There were different realities at work, and that set the stage for those deemed mad in America to suffer in new and novel ways.

7

THE PATIENTS' REALITY

The drugs I had taken for so many months affected every part of my body. My eyes kept going out of focus, especially when I tried to read. My mouth was dry, my tongue swollen, my words slurred. Sometimes I forgot what I was trying to say. My body was puffy. I hadn't menstruated in months and was able to move my bowels only with enormous amounts of laxatives. I had no energy at all. If walking around in a constant haze is supposed to be tranquility, I was successfully tranquilized.

—Judi Chamberlin[1]

THE RECASTING OF neuroleptics, from agents that could help stabilize people suffering from a psychotic episode into safe, antischizophrenic pills, made for a fateful turn in America's care of the "mad." The opportunity at hand in the late 1950s was profound. Eugenic conceptions of the mentally ill had produced a horrible record. The mentally ill had been warehoused in bleak asylums and subjected to such medical treatments as insulin coma, metrazol convulsive therapy, forced electroshock, and lobotomy. With the appointment of the Joint Commission on Mental Illness and Mental Health in 1955, the country had the opportunity to rethink its care of the mentally ill and, equally important, to rethink its conceptions of the mentally ill. Were they biological defectives? Or were they simply people—disturbed in some fashion—who needed to be welcomed back into the human family? The opportunity, in essence, was for the country to rediscover the moral therapy precepts of the Quakers in York and develop a national program of care consistent with their humane conceptions of the "insane."

But once neuroleptics had been refashioned into *antischizophrenic* agents, a very different future was foretold.

Deniker, Delay, Lehmann, and the others who pioneered the use of neuroleptics correctly understood that the drugs achieved their effects not by "normalizing" brain chemistry but by hindering brain function. Precisely how the neuroleptics did so started to become clear in 1963. That year, Swedish pharmacologist Arvid Carlsson determined that neuroleptics inhibit the activity of a chemical messenger in the brain, dopamine. The invention of brain-imaging technologies, such as positron emission tomography, subsequently made it possible to quantify the degree of that inhibition. The relative potency of standard neuroleptics is determined by their affinity for binding the D_2 receptor, which is a particular type of dopamine receptor. At a therapeutic dose, a neuroleptic may occupy 70 percent to 90 percent of all D_2 receptors.[2] With the receptors so blocked, dopamine can't reliably deliver its message to cells. The brain's communication system is thwarted, and any bundle of nerve fibers that relies primarily on D_2 receptors is sharply impaired. That is the mechanism at work with standard neuroleptics. The drugs alter a person's behavior and thinking by partially shutting down vital dopaminergic nerve pathways.

Once that mechanism of action is understood, it becomes clear why neuroleptics produce symptoms similar to Parkinson's disease and also why the drugs provide a type of chemical lobotomy.

There are three prominent dopaminergic pathways in the brain. One, the nigrostriatal system, originates in the basal ganglia and is vital to the initiation and control of motor movement. Parkinson's disease results from the death of dopamine-producing neurons needed to operate this pathway. The patient's brain stops producing an adequate supply of the neurotransmitter—dopamine levels in Parkinson's patients are only about 20 percent of normal—and without it, the pathway malfunctions. Conventional neuroleptics cause Parkinsonism because they produce a similar marked deficiency. Although the patient's brain may still be producing an adequate supply of dopamine, the neurotransmitter is blocked from binding to receptors, and thus the pathway's normal functioning is disrupted. In this manner, neuroleptics can be fairly seen as chemical restraints—they dramatically curb the neurotransmitter activity that underlies motor movement.*

A second dopaminergic pathway, the mesolimbic system, ascends from a midbrain region called the ventral tegmentum to the limbic area. The limbic system, which is located next to the frontal lobes, regulates emotion. It is here that we *feel* the world. This feeling is vital to our sense of self and to our conceptions of reality. From an evolutionary standpoint, it is also designed to be a center for paranoia. It is the limbic system that remains vigilant to environmental dangers, and if danger is seen, it mounts an emotional response. By impairing the limbic system, neuroleptics blunt this arousal response—an effect that has made the drugs useful in veterinary medicine for taming animals. In a similar vein, neuroleptics "tranquilize" people. But for people so tranquilized, this clamping down on the limbic system often translates into an internal landscape in which they feel emotionally cut off from the world. People on neuroleptics complain of feeling like "zombies," their emotions all "wrapped up." In a very real sense, they can no longer emotionally experience themselves.

A third dopaminergic pathway, known as the mesocortical system, ascends from the ventral tegmentum to the frontal lobes.

---

*In 1991, researchers found that a toxin called MPTP, which had been discovered in a batch of contaminated heroin (addicts who injected it became frozen in place), closely resembled three neuroleptics: haloperidol, chlorpromazine, and thiothixene. MPTP was subsequently used to induce Parkinson's in animals so the disease could be studied; haloperidol has been used in such studies as well.

Neuroleptics, by inhibiting this pathway, hinder the communication between these two brain regions. In a like manner, surgical lobotomy involved severing nerve fibers connecting the frontal lobes to the thalamus, another "older" brain region. In both instances, as drug critic Peter Breggin has pointed out, the integration of frontal-lobe function with other brain regions is disrupted.[3] Indeed, experiments with monkeys have shown that if the mesocortical dopaminergic system is impaired, the prefrontal cortex doesn't function well. "Depletion of dopamine in the prefrontal cortex impairs the performance of monkeys in cognitive tasks, similar to the effect of ablating the prefrontal cortex," explains *Principles of Neural Science*, a modern neurology textbook.[4] The frontal lobes rely on dopamine to function, and thus standard neuroleptics, by partially blocking this chemical messenger, provide a kind of pharmacological lobotomy.

What neuroleptics do, then, is induce a *pathological* deficiency in dopamine transmission. They induce, in Deniker's words, a "therapeutic Parkinsonism."[5] And once they became the standard fare in psychiatry, this is the pathology that became the face of madness in America. The image we have today of schizophrenia is not that of madness—whatever that might be—in its natural state. All of the traits that we have come to associate with schizophrenia—the awkward gait, the jerking arm movements, the vacant facial expression, the sleepiness, the lack of initiative—are symptoms due, at least in large part, to a drug-induced deficiency in dopamine transmission. Even behavior that seems contrary to that slothful image, such as the agitated pacing seen in some people with schizophrenia, often arises from neuroleptics. Our perceptions of how those ill with "schizophrenia" think, behave, and look are all perceptions of people altered by medication, and not by any natural course of a "disease."

Grist for the Madness Mill

Once neuroleptics were deemed "antischizophrenic," the presumed medical model at work was straightforward. There was a diagnosable disorder, called schizophrenia, that could be successfully treated with a medication specific to it. That precise correlation of diagnosis and medication even spoke of medicine at its best. An artful diagnosis begat a singularly appropriate treatment. Regardless of the merits of the drugs, it was a model that could be valid only if American psychiatry could reliably diagnose this disorder. But by the 1970s, it became evident that psychiatry had no such skill and that schizophrenia was a term being loosely applied to people with widely disparate emotional problems. It also was a label applied much more quickly to poor people and African-Americans.

The invention of schizophrenia, as a diagnostic term, can be traced back to the work of German psychiatrist Emil Kraepelin. Throughout the nineteenth century, physicians had conjured up a wild profusion of insanity types. Medical texts told of such ailments as "old maid's insanity," "erotomania," "masturbatory psychosis," "pauperism insanity," and "chronic delusional disorder." There was no scientific rhyme or reason to the terms, and they provided little insight into what the future held for the patient. Kraepelin, after studying case histories of asylum patients for more than a decade, put such practices to rest by developing classifications that tied symptoms to predicted outcomes. He divided psychotic disorders into two principal groups. Patients who had psychotic episodes along with emotional disturbances suffered from manic-depressive illness, and they could hope to get better. Psychotic patients who exhibited a lack of affect, or emotion, suffered from dementia praecox (premature dementia). Their predicted fate was much gloomier: Seventy-five percent (or more) could be expected to deteriorate into an end-stage dementia. In 1908, Swiss psychiatrist Eugen Bleuler coined the term "schizophrenia" as a substitute for dementia praecox.

As a result of the work of Kraepelin and Bleuler, twentieth-century psychiatrists have generally held pessimistic views about their schizophrenia patients. The expected poor outcome has also been used to justify aggressive medical treatments. If patients aren't likely to get better, then even brain-disabling treatments like lobotomy might be justified. With schizophrenics, there isn't much to lose. But, as English historian Mary Boyle convincingly argued in 1990, Kraepelin's population of psychotic patients undoubtedly included a number of patients with organic brain diseases, most specifically encephalitis lethargica.[6] In fact, Kraepelin's description of chronic schizophrenics deteriorating over time and sliding into dementia is a description of people stricken by the encephalitis lethargica virus.

In the late 1800s, when Kraepelin was doing his pioneering work, encephalitis lethargica was not a known disease. Anybody suffering from it would have been dumped into the pool of lunatics housed in asylums. This was the patient pool that Kraepelin had tried to sort out, and as he'd done so, he'd identified a common type of patient, which became part of his dementia praecox group, that had peculiar physical symptoms. In addition to their mental and emotional problems, these patients walked oddly and suffered from facial tics, muscle spasms, and sudden bouts of sleepiness. Their pupils reacted sluggishly to light. They also drooled, had difficulty swallowing, were chronically constipated, and were unable to complete willed physical acts. These patients apparently suffered from a global illness, which affected their mental, emotional, and physical spheres, and these were the patients most likely to become demented.

Kraepelin's work was still fresh in physicians' minds when, in the winter of 1916–1917, a mysterious illness broke out in Vienna and other European cities. No one knew quite what to make of the new disease. Those afflicted might suddenly turn delirious, or drop into a stupor, or start walking in a jerky manner. "Epidemic Parkinsonism," "epidemic delirium," and "epidemic schizophrenia" were a few of the phrases used to describe the outbreak, which turned into a worldwide pandemic that lasted until 1927. Very early on, however, Austrian neurologist Constantin von Economo solved the mystery. He found that the brain tissue of dead patients contained an agent (presumably a virus) that could transmit the illness to monkeys. Many also had a characteristic pattern of damage in their brains, most notably in the substantia nigra region (a dopaminergic system in the basal ganglia). He named his infectious disease "encephalitis lethargica."[7]

At the time, the disease was widely seen as "new" to nature. Yet physicians quickly found themselves in a difficult quandary: How could they reliably distinguish it from Kraepelin's schizophrenia? Both von Economo and Kraepelin described their patients' symptoms in very similar terms. Both patient groups suffered muscle spasms, an odd gait, and facial tics. Both suffered from delusions. Both could drop into a profound stupor. And even at autopsy, it seemed that Kraepelin's chronic schizophrenic patients were much like von Economo's. In a number of patients, Kraepelin had microscopically observed severe nerve damage in their brains, along with the proliferation of abnormal glial cells, which was the same kind of damage that von Economo saw in his patients.

Despite the diagnostic confusion, the European medical community remained convinced that the two disorders were distinct. Physicians wrote of subtle features that, at least in theory, could lead to one diagnosis or the other. What few noticed, however, is that once the encephalitis lethargica epidemic waned in the late 1920s, so too did the supply of "schizophrenics" who fit Kraepelin's description of those psychotic patients most likely to have gloomy outcomes. "The inaccessible, the stuporous catatonic, the intellectually deteriorated"—these types of schizophrenia patients, Boyle noted, largely disappeared. The presenting symptoms described by Kraepelin, such as pupillary disorders, dramatic weight loss and gain, and facial tics, were no longer commonly seen.

It is also apparent today that encephalitis lethargica did not make its first appearance in 1917, but long before. In his book *Awakenings,* neurologist Oliver Sacks recounted periodic outbreaks of sleeping sickness dating back at least five centuries. Italy apparently suffered through one in 1889–1890. Psychiatry, however, has unfortunately never gone back to revisit Kraepelin's work. What would he have concluded about psychotic disorders if people ill with encephalitis lethargica had been removed from the asylum patients he'd studied? Would he still have found a group who had no known organic brain pathology but still commonly had poor long-term outcomes? Was his pessimism about schizophrenia justified? Psychiatry never addressed this issue. Schizophrenia was a concept too vital to the profession's claim of medical legitimacy. And so once Kraepelin's deteriorated schizophrenics disappeared, psychiatry simply altered the diagnostic criteria. The physical symptoms of the disease were quietly dropped. The greasy skin, the odd gait, the muscle spasms, the facial tics—all of those symptoms disappeared from the diagnostic manuals. What remained, as the foremost distinguishing features, were the mental symptoms: hallucinations, delusions, and bizarre thoughts. "The referents of schizophrenia," Boyle observed, "gradually changed until the diagnosis came to be applied to a population who bore only a slight, and possibly superficial, resemblance to Kraepelin's."

Thus, the very concept of schizophrenia was born amid diagnostic confusion, and within forty years, it had become something new. In place of the global illness afflicting most of Kraepelin's patients, schizophrenia became a disorder defined primarily by the presence of abnormal thoughts. Once it was so defined, diagnosis naturally became problematic. As William Carpenter, a prominent psychiatrist at the University of Maryland, noted in 1985, delusions and hallucinations are "distortions and exaggerations of normal function."[8] Walter Mitty goes on his walk and fantasizes about being a sports hero. A religious person feels the body of Christ enter her body. Yet another hears the voice of God or that of a long-dead relative. When do such thoughts and voices become pathological, and when are they simply culturally acceptable imaginings? The difficulty in defining this line was dramatized by a 1970s study of 463 people in El Paso, Texas. Researchers found that every single person experienced thoughts, beliefs, moods, and fantasies that, if isolated in a mental health interview, would support a diagnosis of mental illness. The symptoms used to justify a diagnosis of schizophrenia—feelings of being possessed, of extreme paranoia, and of having special powers—were "experienced frequently" by a fair number of people.[9]

Starting in the 1940s, American psychiatrists also began radically altering where they drew the line separating "normal" from "abnormal." Up to that point, only about one-third of patients admitted to New York mental hospitals were diagnosed as schizophrenic. The rest were given less severe diagnoses, like manic-depressive illness. Two decades later, more than half of admitted patients were being diagnosed as schizophrenic. Researchers who compared the diagnostic practices of New York and London psychiatrists found that the American doctors were regularly applying the schizophrenic tag to people who should properly be diagnosed as manic depressive, or even simply neurotic. In one experiment, 69 percent of American psychiatrists shown a video of a socially inept, moody thirty-year-old bachelor diagnosed him as schizophrenic, whereas only 2 percent of the British psychiatrists did. "At least to a European observer," one British psychiatrist concluded in 1971, "the diagnosis is now made so freely on the east coast of the United States that it is losing much of its original meaning."[10]

The liberal use of this diagnosis in the United States arose, at least in part, from underlying political and social tensions. In the 1950s, the Cold War—which pitted the United States, more than any European country, against the Soviet Union—led to a relative lack of tolerance in this country for nonconformist behavior, and that decade gave way to one marked by social protests. There was a clash of cultures, and as this occurred, American psychiatry became ever more quick to judge a person "schizophrenic." Jonika Upton's "symptoms" included carrying Proust under her arm and running off with a boyfriend her parents suspected was a homosexual. Leonard Roy Frank, who became a well-known leader of antipsychiatry protests in the 1970s, was diagnosed as a paranoid schizophrenic in 1962 after he stopped working as a real estate salesman and "dropped out"—he grew a beard, became a vegetarian, and read religious texts. All of these were listed as symptoms of his schizophrenia in his medical records (he was said to be living the "life of a beatnik"), and as a "therapeutic device" physicians even forcibly shaved off his beard.[11]

Numerous studies detailed just how eager American psychiatrists were to make this diagnosis. A researcher who reviewed Manhattan State Hospital's 1982 case records determined that 80 percent of the "schizophrenic" patients there had never exhibited the symptoms necessary to support such a diagnosis. Nationwide, it was estimated in 1978 that more than 100,000 people had been so misdiagnosed. "Psychiatric diagnosis," Canadian psychiatrist Heinz Lehmann scolded his American peers, "in many quarters today has deteriorated from being a fine and useful craft into an ill-regulated, superficial, unconvincing, and therefore often useless procedure."[12]

In 1973, Stanford University psychology professor David Rosenhan memorably proved Lehmann's point. He and seven other "normal" people showed up at twelve different mental hospitals (some went to more than one hospital) complaining that they heard voices, vague in nature, which said such things as "thud," "empty," or "hollow." Those were the only fake symptoms they gave. They behaved calmly and described their relationships with friends and family just as they were. In every instance, the "pseudopatients" were admitted to the hospital, and in every case but one, they were diagnosed as ill with schizophrenia.

Once admitted, they stopped complaining of any symptoms. They even began openly writing in their notebooks, acting as the educated observers they were. In spite of this, none of the hospital staff ever spotted them as impostors. The eight pseudopatients were given 2,100 neuroleptic pills (which they hid or flushed in the toilet, as many of the actual patients did as well). The only ones in the hospital who didn't fall for their ruse were the "real" patients. "You're not crazy," they'd tell the pseudopatients. "You're a journalist, or a professor (referring to their note taking). You're checking up on the hospital." Rosenhan and his colleagues also discovered what it was like to be a schizophrenic in the eyes of others. Doctors and nurses spent almost no time with them, avoided making eye contact, and didn't respond in meaningful ways to even their simplest questions. Often, they were awakened in the morning by attendants screaming, "Come on, you motherfuckers, out of bed."

Rosenhan also ran the experiment in reverse. He told a prestigious teaching hospital that at some point in the following three months, a pseudopatient would attempt to gain admittance to its psychiatric unit. During that ninety-day period, the teaching hospital admitted 193 psychiatric patients, and forty-one were alleged, by at least one member of the staff, to be Rosenhan's impostor. In fact, no pseudopatient had tried to gain admittance. "The facts of the matter are that we have known for a long time that diagnoses are often not useful or reliable, but we have nevertheless continued to use them," Rosenhan wrote in *Science.* "We now know that we cannot distinguish insanity from sanity."[13]

Rosenhan's study was akin to proving that American psychiatry had no clothes. It was evidence that American psychiatry was diagnosing schizophrenia in a willy-nilly, frivolous manner. As if that were not threatening enough, a number of studies showed that American doctors were preferentially applying the label to people with black skin and to the poor.

The diagnosis of mental illness in African-Americans has a shameful history. During the nineteenth century, the perceived mental health of African-Americans was closely tied to their legal status as free men or slaves. Those who lived in free states, or those who were slaves and publicly exhibited a desire to be free, were at particular risk of being seen as insane. According to the 1840 U.S. census, insanity was eleven times more common among Negroes living in the North than in the South. That statistic arose, in part, because whites in some Northern counties reported to census takers that all of the Negroes in their communities were crazy. The 11:1 ratio was quickly shown to be ludicrous, but not before Southern politicians had seized upon it as evidence that bondage was good for Negroes. "Here is proof of the necessity of slavery," reasoned Senator John Calhoun. "The African is incapable of self-care and sinks into lunacy under the burden of freedom. It is a mercy to give him the guardianship and protection from mental death."[14] In 1851, a prominent Southern physician, Samuel Cartwright, took this argument a step further. Writing in the *New Orleans Medical and Surgical Journal,* he said he'd identified two new types of insanity among slaves. One was drapetomania, which was to be diagnosed whenever a Negro sought to run away. He reasoned that slave owners stirred this mental illness by being too kind to "their negroes . . . treating them as equals," which confused the poor slaves because God had made them to be "*submissive knee benders,*" even giving them a superflexible knee joint for this purpose. The other mental disorder he'd discovered was dysaesthesia aethiopis, which was characterized by idleness and improper respect for the master's property. Cartwright advised that light beatings and hard labor reliably cured this mental illness, as such medicine could turn an "arrant rascal" into "a good negro that can hoe or plow."[15]

After the Civil War ended, Southern Negroes, emancipated from their bonds of slavery, found themselves newly at risk of being locked up in mental asylums. The definition of sanity in Negroes was still tied to behavior that a slave owner liked to see: a docile, hardworking laborer who paid him proper respect. Negroes who strayed too far from that behavioral norm were candidates for being declared insane and were put away in asylums, jails, and poorhouses. Nationwide, the incidence of "insanity" among Negroes rose fivefold between 1860 and 1880, and once again, such statistics were seen by many Southern doctors as evidence that the "colored race" simply couldn't handle freedom. Negroes, explained Mississippi asylum physician J. M. Buchanan in 1886, did not have the biological brainpower to live free in a civilized country because "the growth of the [Negro] brain is arrested by premature closing of the cranial sutures." When enslaved, he added, the childish Negro was able to enjoy life, "fat, sleek, and contented," his mind unburdened by cares, and "his passions and animal instincts kept in abeyance by the will of his master."[16] Thirty-five years later, W. M. Bevis, a physician at St. Elizabeth's Hospital in Washington, D.C., revisited this theory in the *American Journal of Psychiatry*. Negroes were particularly prone to psychotic illness, he wrote, because they were descendants of "savages and cannibals" and thus, as free men in America, were living in "an environment of higher civilization for which the biological development of the race had not made adequate preparation." All of this led one African-American scholar, E. Franklin Frazier, to suggest in 1927 that perhaps whites who were racially prejudiced and acted cruelly toward blacks (mob violence, lynchings, and so on) should be seen as insane, a viewpoint that got him fired from his post as director of the Atlanta University School of Social Work. So great was the furor that Frazier, armed with a gun for self-protection, fled the city at night.[17]

In the first part of the twentieth century, the funneling of blacks into the schizophrenic category, as opposed to their being given a diagnosis of manic-depressive insanity or involutional melancholy, was also due to cultural beliefs that blacks were happy-go-lucky and lacked the intelligence to worry about the myriad stresses in life. They might become maniacal or crazy in their thoughts but—or so the belief went—they weren't very likely to become morbidly sad. "Depressions of various forms are rare in the colored," explained Mary O'Malley, a physician at St. Elizabeth's Hospital. "These individuals do not react to the graver emotions—grief, remorse, etc.—owing to the fact that they have no strict moral standard and no scrupulosity as to social conventions."[18] Although that happy-go-lucky stereotype may have dissipated in the second half of the century, the funneling of blacks into the schizophrenic category did not. A 1982 study of 1,023 African-Americans said to be schizophrenic determined that 64 percent didn't exhibit symptoms necessary, under prevailing American Psychiatric Association (APA) guidelines, for making such a diagnosis. Other studies found that blacks were being preferentially put into subcategories of schizophrenia that "connote dangerousness and (pathological) severity," and that in comparison with whites, they were more likely to be committed against their will to a psychiatric unit. A 1988 experiment by two sociologists at Indiana University, Marti Loring and Brian Powell, revealed just how deeply ingrained the bias is. They had 290 psychiatrists review written case studies in which the patients were alternatively described as white male, white female, black male, and black female (but otherwise the details remained the same). The psychiatrists' diagnoses diverged in two directions from the norm: More severe for black males and less severe for white males. Wrote Loring and Powell: "Clinicians appear to ascribe violence, suspiciousness, and dangerousness to black clients even though the case studies are the same as the case studies for the white clients."[19]

The overrepresentation of the poor among the "insane" is an old story in American psychiatry. To a large degree, the crowded asylums in the nineteenth century served as poorhouses. They were filled with social misfits, the chronically ill, and the emotionally troubled—"insanity" was simply a legal term for confining this diverse group. The one thing that nearly all patients in municipal asylums did have in common was that they were paupers. Edward Jarvis, in his 1855 report on insanity in Massachusetts, calculated that "insanity" was sixty-four times more common among the financially destitute than among the rest of the population.[20] One hundred thirty years later, epidemiologists reported that the poverty link still held true: People in the bottom quartile of the socioeconomic ladder had nearly eight times the risk of being diagnosed schizophrenic as people from the top quartile.[21] Behaviors and emotions that can lead to a diagnosis of schizophrenia—hostility, anger, emotional withdrawal, paranoia—go hand in hand with being poor.

All of this—Rosenhan's experiment, the divergence in diagnostic practices between American and English doctors, the preferential labeling of blacks and the poor—point to one inescapable conclusion. As has often been observed, good medicine begins with an artful diagnosis. But during the 1960s and 1970s, something akin to the reverse of that became the norm in American psychiatry. People with widely disparate emotional and behavior problems—some anxious, some morbidly depressed, some hostile, and some afflicted with odd notions and bizarre thoughts—were regularly funneled into a single diagnostic category, schizophrenia, and then treated with neuroleptics. At that point, their behavior and underlying brain chemistry did become more alike. They would now all show evidence of a drug-induced deficiency in dopamine transmission. And with the schizophrenia label applied, others would treat them—"Come on, you motherfuckers, out of bed"—in ways that confirmed their new medical status. American medicine, in essence, had developed a process for minting "schizophrenics" from a troubled cast of people, with blacks and the poor most at risk of being so transformed.

In 1985, Alan Lipton, chief of psychiatric services for New York state, detailed the manufacturing process at work. He reviewed the case histories of eighty-nine patients at Manhattan State Hospital who had been diagnosed as schizophrenic, and found that only sixteen, based on their initial symptoms, should have been so classified. But then the medical process took over:

> The self-fulfilling prognostic prophecy of gloom in schizophrenia . . . was painfully evident in the histories of most of our patients. Most often, once written, the diagnosis of schizophrenia became irrevocable and apparently was never reconsidered. The probability of such reconsideration was further lessened by the effects of the inevitable neuroleptics, prescribed in doses sufficient to "quiet" the "disturbing" symptoms. Since manic disorders respond symptomatically to sufficient neuroleptic medications, and even major depressions with or without mood incongruent delusions can be suppressed by these drugs, a relatively homogenous population of "medicated schizophrenics" has been created. The subsequent adaptation of this population to institutional and social demands reinforces and eventually congeals their homogeneity.[22]

In sum, diagnosis begat the disease of "medicated schizophrenic."

By revisiting Kraepelin's early work, one can also foresee the expected outcomes for such drug-altered patients. The very symptoms that, in Kraepelin's time, predicted the worst outcomes in psychotic patients—an odd gait, muscle spasms, extreme lethargy, and facial twitches—all *reappeared* in the schizophrenic population once neuroleptic medications were introduced. Encephalitis lethargica damaged dopaminergic systems in the brain. Neuroleptics, by partially shutting down dopaminergic transmission, created a similar biological pathology. Modern medication brought back to life the very class of psychotic patients that Kraepelin had identified as most likely to become chronically ill and to deteriorate into dementia.

The Patient's Point of View

The evaluation of the merits of medical treatments for madness has always been a calculation made by doctors and, to a certain extent, by society as a whole. Does the treatment provide a method for managing disturbed people? That is the usual bottom line. The patient's subjective response to the treatment—does it help the patient feel better or think more clearly?—simply doesn't count in that evaluation. The "mad," in fact, are dismissed as unreliable witnesses. How can a person crazy in mind possibly appreciate whether a treatment—be it Rush's gyrator, a wet pack, gastrointestinal surgery, metrazol convulsive therapy, electroshock, or a neuroleptic—has helped? Yet to the person so treated, the subjective experience is everything.

The "mad," being a diverse lot, responded in varying ways to neuroleptics. The drugs themselves became somewhat varied in kind, and ever more potent. Thorazine eventually gave way as the neuroleptic of choice to Prolixin (fluphenazine), a long-acting neuroleptic that could be injected, and to Haldol (haloperidol)—and these latter two drugs clamped down on dopamine transmission in a more robust manner. With this variability both in patients and in drugs, subjective responses to neuroleptics were unpredictable. Some patients experienced the drug-induced change in brain function as a positive, reporting that the drugs made them calmer, less fearful, and even clearer in mind. (Or at least they told their psychiatrists that—there are noticeably few writings by ex-patients in praise of standard neuroleptics.) The much more common response by patients, however, was decidedly negative. Patients complained that the drugs turned them into "zombies" or made them feel "closed in," "mummified," "jittery," "confused," and "fearful." They described their medications as "poisons" that produced "the worst misery."[23] Ex-patients wrote of the drugs as the latest form of "psychiatric assault."

One who so described her experiences was Janet Gotkin. In the early 1960s, during her first year at college, she became distraught and suicidal. Over the course of the next ten years, she was prescribed more than 1 million milligrams of neuroleptics, often at fantastically high doses (up to 2,000 milligrams of Thorazine a day, ten times what physicians in the early 1950s described as producing a chemical lobotomy). In her book *Too Much Anger, Too Many Tears,* and in testimony at a 1975 Senate hearing, she told of how the drugs "turned me into a fucking invalid, all in the name of mental health."

> I became alienated from my self, my thoughts, my life, a stranger in the normal world, a prisoner of drugs and psychiatric mystification, unable to survive anywhere but in a psychiatric hospital. The anxieties and fears I had lay encased in a Thorazine cocoon and my body, heavy as a bear's, lumbered and lurched as I tried to maneuver the curves of my outside world. My tongue was so fuzzy, so thick, I could barely speak. Always I needed water and even with it my loose tongue often could not shape the words. It was so hard to think, the effort was so great; more often than not I would fall into a stupor of not caring or I would go to sleep. In eight years I did not read an entire book, a newspaper, or see a whole movie. I could not focus my blurred eyes to read and I always fell asleep at a film. People's voices came through filtered, strange. They could not penetrate my Thorazine fog; and I could not escape my drug prison. The drugs made me constipated as well as ravenously hungry. As a final misery, they caused me to gain weight. For eight years, I took laxatives and suffered as I watched my body grow heavy and distorted. My hands shook so I could barely hold a pencil and I was afflicted with what Dr. Sternfield lightly called "dancing legs," a Parkinsonian "side effect" of these chemicals. For this I took a drug called Kemadrin, and if I missed a day or a dosage, my shoulder

> muscles would tighten into excruciatingly painful knots and my legs would go on wildly out of control. . . . These drugs are used, not to heal or help, but to torture and control. It is that simple.[24]

The Senate subcommittee that Gotkin told her story to was chaired by Indiana's Birch Bayh, and it was investigating the use of neuroleptics in juvenile institutions, primarily jails and homes for the retarded. But it was ex-mental patients, in oral testimony and in writings that were made part of the public record, who told the legislators what it was truly like to be on these drugs. "After the first few injections I had a very common physical reaction to the drug," wrote Daniel Eisenberg. "My mouth became locked and frozen in an open position in excruciating pain." Another ex-patient, Beth Guiros, described her newfound shame: "I was so heavily drugged that I had difficulty in walking or talking . . . I was nicknamed zombie. I was told how gross I looked and to shut my mouth and quit drooling." The drugs led Anil Fahini to "the most fatalistic and despairing moments I've had on this planet. The only way I can describe the despair is that my consciousness was being beaten back . . . They prevent you from carrying on thought processes. They hold you in a tight circle of thoughts that never find fulfillment, that never find freedom of expression." Wade Hudson, a graduate of the University of California at Berkeley, told Bayh what it was like to be injected with Prolixin:

> After ten days or so, the effects of the Prolixin began building up in my system and my body started going through pure hell. It is very hard to describe the effects of this drug and others like it. That is why we use strange words like zombie. In my case, the experience became sheer torture. Different muscles began twitching uncontrollably. My mouth was like very dry cotton no matter how much water I drank. My tongue became all swollen up. My entire body felt like it was being twisted up in contortions inside by some unseen wringer. And my mind became clouded up and slowed down. Before, I had been reasoning incorrectly but at least I could reason. But most disturbing of all was that I feared that all of these excruciating [drug-induced] experiences were in my mind, or caused by my mind, a sign of my supposed wickedness.[25]

Physical pain, an inability to reason, alienation from the self, flattened emotions toward others, inner despair—these themes were constant in the patients' stories. In his 1992 book *How to Become a Schizophrenic*, John Modrow summed up his experience this way:

> I was totally unable to take those drugs without constantly reminding myself that I was a *schizophrenic*—a pitiful, helpless *defective* human being . . . taking neuroleptic drugs causes a loss of self-esteem, a sense of helplessness and hopelessness, and a total paralysis of will. While taking those drugs it is nearly impossible to view oneself as a free agent. In taking those drugs one is being conditioned to see oneself as a defective object subject to forces totally beyond one's control.[26]

This subjective experience was not unique to the mad. The not-so-mad recounted the same thing. Psychiatrist Nathaniel Lehrman, who was clinical director of Kingsboro Psychiatric Center in Brooklyn from 1973 to 1978, was treated with Thorazine in 1963, after suffering a psychotic break that landed him in a mental hospital. He quickly began "tonguing" the medication, hiding it in his cheek and then spitting it out. This was critical in his recovery, Lehrman said, in an interview.

> The [psychotic] break itself was unpleasant. The effects of the medication were even more unpleasant. At first, the Thorazine seemed to speed up my metabolic processes. The next day, there was an unhinging between my thoughts and my feelings. My feelings were grossly disproportional to my thoughts. I thought the only way to survive was wrapped in swaddling clothes. I couldn't stand up straight. My eyes weren't focusing properly, and walking—or anything else, even thinking—became a terrible effort. I couldn't even read. The medication was robbing me of my will, and of any control I had over my own fate. I got better by running a mile each day, playing my violin, and starting up a research study. These are activities that are useful and satisfying, and that's what people need. I couldn't have done those things if I had stayed on the medication.[27]

Two Israeli physicians, Robert Belmaker and David Wald, writing in the *British Journal of Psychiatry*, told how a single dose of Haldol knocked them for a loop:

> The effect was marked and very similar in both of us: within ten minutes a marked slowing of thinking and movement developed, along with profound inner restlessness. Neither subject [the two doctors] could continue work, and each left work for over 36 hours. Each subject complained of a paralysis of volition, a lack of physical and psychic energy. The subjects felt unable to read, telephone or perform household tasks of their own will, but could perform these tasks if demanded to do so. There was no sleepiness or sedation; on the contrary, both subjects complained of severe anxiety.[28]

What all this attests to is that standard neuroleptics, by dampening down the dopamine system, produce a profound change. At the heart of the subjective experience is this: The drugs rob from a person a full sense of *being*. A person on neuroleptics can no longer fully feel the outside world, can no longer fully feel his or her own mind, and can no longer think well. As Marjorie Wallace, a British journalist who helped establish a telephone hot line for the mentally ill, said, what medicated patients often most missed was *themselves*.

> Why do so many of our callers refuse to take or resent taking their medication? We find that, in the anonymity of phone calls to SANE-LINE, even the most deluded person is often extraordinarily articulate and lucid on the subject of their medication; they know the names, the spellings of the drugs and the dosage and they can report the side effects quite objectively. "When I take my medication, I feel as though I am walking with lead in my shoes" one young man told me on the telephone. Another told the volunteer who took his call, "I feel emptied out, devoid of ideas." Another young man sent us a poem in which he compares the effect of the drugs with drowning—"I was always under the water gasping for air and sunshine," he writes . . . Almost all of our callers report sensations of being separated from the outside world by a glass screen, that their senses are

> numbed, their willpower drained and their lives meaningless. It is these insidious effects that appear to trouble our callers much more than the dramatic physical ones, such as muscular spasms.[29]

A young man interviewed by Wallace put it even more poetically: "I want to stop my medication. I want to dream my own dreams, however frightening they may be, and not waste my life in the empty sleep of drugs."

The 1975 Senate hearings seemingly brought this complaint of mental patients to a sharp, public focus. By that time, neuroleptics were routinely being used in institutions housing the elderly, juvenile delinquents, and the retarded, and at the hearings, a long line of social workers, lawyers, and youth advocates denounced such drugging as punishment of the most unusual and cruel sort. People so medicated, said one witness, "suffer a new and deadlier confinement" than prisoners had ever known in the past. Neuroleptics, said another, "rob you of your mind, your dignity, and maybe your life." Bayh summed up his feelings with similar rhetoric, calling neuroleptics "chemical handcuffs" that assured "solitary confinement of the mind." This was a powerful chorus of opinion that seemingly put American psychiatry on the hot spot. How could psychiatry possibly justify giving such drugs, denounced in a Senate hearing as instruments of mental torture, to the mentally ill? But then Bayh, in a most remarkable statement, carved out one exception to the general rule. "We are not concerned about those [medical] situations where those drugs are used appropriately after proper diagnosis."[30]

At that point, the mentally ill had been put neatly into a box separate from the rest of humanity. What everyone else experienced as mental handcuffs, as a form of chemically imposed solitary confinement, was a proper medical treatment for those said to be "schizophrenic."

A Pathology Defined

An understanding of any brain pathology is captured in three parts: the presenting symptoms, the individual's subjective experience of those symptoms, and the course of the disorder over time.

Neuroleptics, of course, caused a pathological disruption in dopamine transmission, and gradually an understanding of this pathology came together in that same three-part fashion. The presenting symptoms caused by neuroleptics had been well identified in the 1950s: Patients became lethargic, retarded in movement, and emotionally indifferent. After the Bayh hearing, mental patients had stepped forward to make their subjective experiences known. The third aspect of this "disease" process—how neuroleptics affected people over time—took longer to flesh out. But by the mid-1980s, a fairly clear profile of the long-term course of "medicated schizophrenia" had emerged in the medical literature. The drugs made people chronically ill, more prone to violence and criminal behavior, and more socially withdrawn. Permanent brain damage and early death were two other consequences of neuroleptic use.

Ever More Crazy

Evidence that neuroleptics were making people chronically ill showed up fairly early. In 1967, NIMH investigators reported on one-year outcomes for the 270 patients in its earlier six-week study that had declared neuroleptics to be antischizophrenic drugs. Much to their surprise, the patients that had not been treated in the hospitals with drugs "were *less* likely to be rehospitalized than those who received any of the three active phenothiazines." The researchers, scrambling to come up with an explanation for this finding, speculated that perhaps that perhaps hospital staff during the initial trial had felt sorry for the placebo patients (because they weren't getting well as fast as the drug-treated patients) and thus had given them "some special quality in care, treatment, or concern" that led to the better one-year outcomes.[31] It was an explanation that revealed more about the researchers than the patients: The NIMH investigators simply couldn't conceive of the possibility that neuroleptics were harming people.

Four years later, however, NIMH physicians were back with another disturbing finding. In a twenty-four-week drug-withdrawal study involving 301 patients, relapse rates *rose* in direct correlation to initial drug dosage, and the no-dosage group had by far the lowest relapse rate.* Only 7 percent of patients who weren't medicated at the start of the study relapsed, compared to 45 percent who were placed on neuroleptics and then had their drugs withdrawn. Moreover, the higher the neuroleptic dosage that patients were on before withdrawal, the higher the relapse rate.[32]

Something clearly was amiss. Both of these studies suggested that neuroleptics altered brain physiology in a way that made people more biologically prone to psychosis. Other reports soon deepened this suspicion. Even when patients reliably took their medications, relapse was common, and researchers reported in 1976 that it appeared that "relapse during drug administration is greater in severity than when no drugs are given." Relapsing while on long-acting Prolixin (fluphenazine) was even worse. That led to "severe clinical deterioration." Similarly, if one relapsed while going abruptly off the drugs, psychotic symptoms tended "to persist and intensify." Neuroleptics weren't just making people more vulnerable to relapse. They were also making those who relapsed sicker than they otherwise would have been.[33]

There was one other relapse-related problem with neuroleptics. Often, people going off neuroleptics experienced agonizing withdrawal symptoms, which made it that much more difficult for them to return to a drug-free state. Nausea, diarrhea, headaches, anxiety, insomnia, "rebound" psychosis, and muscular spasms were commonplace. Sol Morris, who lives in Oregon, went through it multiple times, starting when he was fifteen years old:

> It's a nightmare. And it's not just mental pain. The physical pain is indescribable. You can't urinate but you feel like you have to urinate, then you finally can urinate and you feel like fire is coming out of you. You feel like you have to sleep but you can't. You shake all the time. I felt like there were fire ants that had got underneath my skin and were biting me all the time. I'd be up in the middle of

*Drug studies in schizophrenia regularly use "relapse" as an outcome measurement. However, what constitutes relapse isn't well defined. Some investigators use rehospitalization as the criteria for relapse. Others define it simply as some degree of worsening—an increase in "psychotic" thoughts or agitated behavior. From a scientific standpoint, it's a loose term at best.

> the night, trembling inside and shaking, scratching myself. I'd be going nuts, thinking I was losing my mind. You feel so alone and isolated from the rest of the world. And so then I'd start taking the drugs again. The drugs just screw everything up.[34]

All of this led at least a few investigators to rethink this drug-based model of care. What, they wondered, would relapse rates be for schizophrenics who were never exposed to neuroleptics? Would people treated with social support, but no drugs, have better outcomes? Boston psychiatrists J. Sanbourne Bockoven and Harry Solomon sought to answer the question by digging into the past. They found that 45 percent of patients treated at Boston Psychopathic Hospital in 1947, with a progressive model of care that emphasized community support, did not relapse in the five years following discharge, and that 76 percent were successfully living in the community at the end of that follow-up period. In contrast, only 31 percent of patients treated in 1967 with drugs at a community health center remained relapse-free over the next five years, and as a group they were much more "socially dependent"—on welfare and needing other forms of support—than those in the 1947 cohort. Three other research groups launched experiments to study the question and came up with similar results. At the NIMH, William Carpenter and other investigators randomized forty-nine schizophrenia patients into drug and nondrug cohorts and provided all of them with psychosocial support. In 1977, they reported that only 35 percent of the nonmedicated patients had relapsed within a year after discharge, compared to 45 percent of those treated with neuroleptics. The medicated patients also suffered more from depression, blunted emotions, and retarded movements. A year later, Maurice Rappaport and his San Francisco colleagues reported that in a randomized trial of eighty young male schizophrenics admitted to a state hospital, only 27 percent of patients treated without neuroleptics relapsed in the three years following discharge, compared to 62 percent of the medicated group. The final study came from Loren Mosher, head of schizophrenia research at the NIMH. In 1979, he reported that patients who were treated without neuroleptics in an experimental home staffed by nonprofessionals had lower relapse rates over a two-year period than a control group treated conventionally in a hospital. As in the other studies, Mosher reported that the patients treated without drugs were the better functioning group as well.[35]

The studies all pointed to the same conclusion: Exposure to neuroleptics increased the long-term incidence of relapse. Carpenter's group defined the conundrum:

> There is no question that, once patients are placed on medication, they are less vulnerable to relapse if maintained on neuroleptics. But what if these patients had never been treated with drugs to begin with? . . . We raise the possibility that antipsychotic medication may make some schizophrenic patients more vulnerable to future relapse than would be the case in the natural course of their illness.[36]

In the late 1970s, two physicians at McGill University in Montreal, Guy Chouinard and Barry Jones, offered a biological explanation for why this was so. The brain responds to neuroleptics—the blocking of dopamine transmission—as though it were a pathological insult. To compensate, dopaminergic brain cells sprout more D_2 receptors. The density of such receptors may increase by more than 50 percent. At first, this compensatory mechanism may alleviate some of the physical and emotional deficits caused by neuroleptics. Parkinson's symptoms may diminish and the extreme emotional lethargy may lift. But the brain is now physiologically changed. It is now "supersensitive" to dopamine, and this neurotransmitter is thought to be a mediator of psychosis. The person has become more biologically vulnerable to psychosis and is at particularly high risk of severe relapse should he or she abruptly quit taking the drugs. As Jones bluntly put it at a 1979 meeting of the Canadian Psychiatric Association: "Some patients who seem to require lifelong neuroleptics may actually do so because of this therapy."[37]

It was also apparent that the shift in outcomes due to neuroleptic use—away from recovery and toward chronic illness—was a pronounced one. Bockoven's study and the other experiments all suggested that with minimal or no exposure to neuroleptics, perhaps 50 percent of people who suffered a psychotic break and were diagnosed with schizophrenia wouldn't

TABLE 7.1
Stay-Well Rates for Patients Treated Without Neuroleptics

Researcher	Treatment Years	No. of Patients	Study Period (in years)	Never Relapsed
Bockoven, 1972	1833–1846	1,173	40+	48%
Lehrman, 1960	1943–44	2,941	5	44%
Bockoven, 1975	1947	100	5	45%
Rachlin, 1956	1950	317	4	52%
Carpenter, 1977	1970s	27	1	65%
Rappaport, 1978	1970s	41	3	73%

SOURCES: J. Sanbourne Bockoven, *Moral Treatment in American Psychiatry* (Springer Publishing, 1972); Nathaniel Lehrman, "A State Hospital Population Five Years After Admission," *Psychiatric Quarterly* 34 (1960):658–681; H. L. Rachlin, "Follow-Up Study of 317 Patients discharged from Hillside Hospital in 1950," *Journal of Hillside Hospital* 5 (1956):17–40; J. Sanbourne Bockoven, "Comparison of Two Five-Year Follow-Up Studies: 1947 to 1952 and 1967 to 1972," *American Journal of Psychiatry* 132 (1975):796–801; William Carpenter, Jr., "The Treatment of Acute Schizophrenia Without Drugs," *American Journal of Psychiatry* 134 (1977):14–20; and Maurice Rappaport, "Are There Schizophrenics for Whom Drugs May Be Unnecessary or Contraindicated?" *International Pharmacopsychiatry* 13 (1978):100–111.

relapse after leaving the hospital, and as many as 75 percent would function fairly well over the long term.[38] The long-term course of the disorder would be fairly benign for the majority of patients, and they wouldn't suffer all the cognitive, emotional, and physical deficits imposed by neuroleptics. They would have *real* lives. However, once "first-episode" patients were treated with neuroleptics, a far different fate awaited them. If they relapsed while on the medication, which 40 percent did in the first year, they would likely descend into a deeper psychosis than they had known before. If they abruptly stopped taking their medication, which many did, they would likely suffer intense withdrawal symptoms, and they would be at much higher risk of relapsing than if they had never been exposed to the drugs. The use of neuroleptics diminished the possibility that a person, distraught in mind and soul when first treated, could ever return to a healthy, nonmedicated life.*

The Madman of Our Nightmares

Supersensitive psychosis was evidence that dampening down the dopamine system could produce paradoxical effects. Neuroleptics temporarily dimmed psychosis but over the long run made patients more biologically prone to it. A second paradoxical effect, one that cropped up with the more potent neuroleptics, particularly Prolixin and Haldol, was a side effect called akathisia. Neuroleptics were supposed to tranquilize patients, but the more powerful drugs often triggered extreme inner anxiety and restlessness. Patients would endlessly pace, fidget in their chairs, and wring their hands—actions that reflected an inner torment. This side effect was also linked to assaultive, violent behavior.

Although the public may think that "crazy" people are likely to behave in violent ways, this was not true of mental patients prior to the introduction of neuroleptics. Before 1955, four studies found that patients discharged from mental hospitals committed crimes at either the same or a lower rate than the general population. However, eight studies conducted from 1965 to 1979 determined that discharged patients were being arrested at rates that exceeded those of the general population.[40] And while there may have been many social causes for this change in relative arrest rates (homelessness among the

---

*In 1995, Peter Weiden, a psychiatrist at St. Luke's–Roosevelt Hospital in New York City, tallied up how all this plays out in the "real" world. Eighty percent of schizophrenia patients treated with standard neuroleptics relapse within two years of hospital discharge, and the majority of the relapsers become sick again while reliably taking their medications. The American Psychiatric Association, in its 1980 diagnostic manual, even described this common downward spiral into chronic illness: "The most common course [of schizophrenia] is one of acute exacerbations with increasing residual impairments between episodes . . . A complete return to premorbid functioning is unusual, so rare, in fact, that some clinicians would question the diagnosis." That gloomy prognosis does not fit the spectrum of outcomes natural to schizophrenia, but it does accurately describe outcomes as shaped by standard neuroleptics.[39]

mentally ill is an obvious cause), akathisia was also clearly a contributing factor.

In his book *In the Belly of the Beast,* Jack Henry Abbott described how akathisia could turn one inside out:

> These drugs, in this family, do not calm or sedate the nerves. They attack. They attack from so deep inside you, you cannot locate the source of the pain . . . The muscles of your jawbone go berserk, so that you bite the inside of your mouth and your jaw locks and the pain throbs. For *hours* every day this will occur. Your spinal column stiffens so that you can hardly move your head or your neck and sometimes your back bends like a bow and you cannot stand up. The pain *grinds* into your *fiber* . . . You ache with restlessness, so you feel you have to walk, to pace. And then as soon as you start pacing, the opposite occurs to you; you must sit and rest. Back and forth, up and down you go in pain you cannot locate, in such wretched anxiety you are overwhelmed, because you cannot get relief even in *breathing.*[41]

Akathisia was given little attention by psychiatric researchers for nearly twenty years. Physicians usually perceived the restless behavior as a sign that the patient was about to relapse and would increase the dosage of the offending drug. But when investigators finally studied it, patients gave them an earful. They told of pain so great that they wanted to "jump out of their skins," of "anxiety of annihilating proportions." One woman banged her head against the wall and cried, "I just want to get rid of this whole body!" Case studies detailed how patients, seeking to escape from this misery, had jumped from buildings, hung themselves, and stabbed themselves. In one study, 79 percent of mentally ill patients who had tried to kill themselves suffered from akathisia. Another study documented thirty cases of akathisia-linked suicides. "They made many requests or demands that something be done to relieve their tensions," the researchers said. "They appeared driven to find some kind of relief." One who killed himself for this reason was a thirty-six-year-old Hispanic man who'd come to a hospital because he couldn't sleep and was overly nervous. He was given an injection of long-acting fluphenazine, and then, over the next several weeks, he repeatedly returned to hospital emergency rooms in an extremely agitated state and "begged for help." Something had to be done about the extreme physical misery he was in, but nothing was, and finally "he killed himself without warning by jumping in front of a subway train." UCLA psychiatrist Theodore van Putten determined that 75 percent of patients treated with a Haldol injection experienced akathisia.[42]

Various investigators found that this side effect regularly made patients more prone to violence. A 1990 study determined that 50 percent of all fights on a psychiatric ward could be tied to akathisia. Yet another concluded that moderate-to-high doses of haloperidol made half of the patients markedly more aggressive. Patients described "violent urges to assault anyone near" and wanting to kill "the motherfuckers" tormenting them in this way. A few case reports linked akathisia to bizarre murders. One thirty-nine-year-old white man—after a haloperidol injection made him feel like he was "falling apart, that . . . all the bones in his body were broken"—bludgeoned his mother with a hammer, an act he later found incomprehensible. Another thirty-five-year-old man, asked why he had stabbed a grocer he had known for some time, said he did it to get the drug-induced pain out of his head: "The only reason I knifed the guy was Haldol messed me up. Prolixin makes me want to kill, too." The murderous explosion of a twenty-three-year-old man, detailed in the *Journal of Forensic Psychiatry*, was perhaps the most chilling example of all. After his wife left him, he became distraught and was brought to an emergency room by the police. He had been briefly hospitalized a number of times before, and he warned the staff that he reacted badly to haloperidol. In spite of his protest, he was injected with the drug, and he quickly exploded in rage. He ran from the emergency room, tore off his clothes in a nearby park, and started attacking everyone he saw. Over the course of forty-five minutes, he tried to rape a woman walking in the park, broke into a house and beat an eighty-one-year-old woman to a pulp, fought with a policeman and then escaped, stabbed two more women, and was then at last subdued by a gang of eight cops.[43]

Such case reports led researchers to conclude that haloperidol could produce a "marked increase in violent behavior," even among those without any history of assault. They dubbed this side effect to neuroleptics a "behavioral toxicity." Little could the public have suspected that the madman of its nightmares, who kills without warning and for no apparent reason, was not always driven by an evil within but rather by a popular medication.*

A Life Alone

Chronic illness, assaultive behavior—these were two disturbing results arising from neuroleptic use. A third disturbing result was found, ironically, in "good outcome" patients. In 1980, NIMH researchers concluded that patients who successfully "stabilized" on neuroleptics became ever more socially withdrawn. However, much as they had done in 1967 when they explained away the low rehospitalization rates among placebo patients, they didn't blame the drugs for this social isolation. Instead, they speculated that it was a survival strategy chosen by the patients:

> The finding in the present study that nonrelapsed patients show an increase in both blunted affect and emotional withdrawal [over time] could be interpreted as indicating that these patients are successful because they can maintain a degree of withdrawal, rather than despite it. It may be that the social withdrawal that characterizes the marginally adjusted schizophrenic patient either in the hospital or in the community is the mechanism whereby that fragile adjustment is maintained.[44]

---

*The fact that patients who abruptly go off neuroleptics may become more wildly psychotic than they otherwise ever would have been is another reason that neuroleptic use may make the mentally ill more prone to violence. Several high-profile murders in recent years have been committed by people in this drug-withdrawal state. Most recently, *Newsweek* reported that Andrea Yates, the Houston mother who killed her five children, did so after "she was taken off the powerful anti-psychotic drug Haldol." However, such instances of violent murders are inevitably reported as examples of why the mentally ill need to be kept on medications, rather than as examples of the peril of using the drugs in the first place. The blame is put on the patients and their "disease," rather than on the medications.

In 1979, Theodore van Putten described how that "success" translated into real life. In a study of forty-six stable, medicated schizophrenics living at thirteen boarding homes, he found that they spent their waking hours in "virtual solitude, either staring vacantly at television (few residents reported having a favorite television show; most were puzzled at the question), or wandering aimlessly around the neighborhood, sometimes stopping for a nap on a lawn or a park bench." Few had hobbies, few worked, and many couldn't even bathe by themselves. The caretakers at the boarding homes treated them like "incompetent, childlike persons." Van Putten concluded: "They are bland, passive, lack initiative, have blunted affect, make short, laconic replies to direct questions, and do not volunteer symptoms . . . there is a lack not only of interaction [with others] and initiative, but of any activity whatsoever."[45]

In short, investigators determined that people who tolerated neuroleptics well, and weren't "relapsing," were living purposeless, isolated, and largely friendless existences.

A Progressive Brain Disease

Neuroleptics, by dampening down the dopamine system, produce an immediate pathology in brain function. By 1959, a case report had appeared in the literature suggesting that the drugs could also cause permanent brain damage—even if the drugs were withdrawn, motor dysfunction remained. That year, French psychiatrists reported the bizarre symptoms that came to be known as tardive dyskinesia (TD): "The tongue [is] permanently projected forward and backward following a rapid rhythm; at times the projection is to the side, sometimes to the right, sometimes to the left . . . the lips participate in this dyskinesia in the form of stereotyped suction motions, pursing, rolling and incessant champing in synergy with rhythmic contractions of the jaw."[46]

This was a description that clearly indicated something had gone horribly awry with the brain center controlling motor movement. It also soon became clear that this disorder did not only affect the facial muscles. People suffered from jerky, spasmodic motions of all types. Arms, ankles, fingers, toes, torso, neck, and larynx could all be affected. Some patients had difficulty walking, sitting, or standing. At times, their speech became incomprehensible, and they had so much trouble swallowing that eating became problematic. In its more severe forms, noted NIMH physician George Crane, TD resembled "in every respect known neurological diseases, such as Huntington's disease, dystonia musculorum deformans, and postencephalitic brain damage." TD was found to appear in 5 percent of patients within one year of neuroleptic treatment, with the percentage so afflicted increasing an additional 5 percent with each additional year of exposure.[47]

Although TD came to be thought of primarily as a motor disorder, dopamine systems are also involved in intellectual and behavioral functions, and thus, as would be expected, patients with tardive dyskinesia are often impaired in these realms as well. Many patients with tardive dyskinesia show accelerated impairment in learning, memory, and a variety of other intellectual tasks. More than twenty studies have documented such deficits. In one study, 44 percent of tardive dyskinesia patients weren't even aware of their motor dysfunction, evidence that they had lost the capacity of mind to monitor their own physical well-being. As one researcher concluded, the weird tongue movements common to tardive dyskinesia may warn of a "larval dementia."[48]

The neuropathology underlying this brain dysfunction is still not well understood. Neuroleptics have been found to cause a dizzying array of pathological changes in the brain. One thought is that the drugs damage the basal ganglia in direct ways. In rats, neuroleptics have been shown to cause a loss of cells in this brain region. Autopsy and magnetic resonance imaging (MRI) studies have also found lesions in the basal ganglia of some TD patients, leading researchers to compare TD to the degenerative processes characteristic of Parkinson's and Huntington's diseases. Harvard scientists have argued that neuroleptics damage neurons because they "elevate levels of oxidative stress."[49]

Researchers have also speculated that TD is related to the brain becoming supersensitive to dopamine. Imaging studies have found that neuroleptic use is associated with hypertrophy of the thalamus and basal ganglion structures (caudate nucleus, putamen, and globus pallidus), and it is thought that this pathological enlargement is the result of an increase in dopamine receptors. In addition to being a possible cause of TD, this hypertrophy, which can be seen with MRIs within eighteen months of neuroleptic use, was found by University of Pennsylvania investigators to be "associated with greater severity of both negative and positive symptoms." In essence, this study provided direct visual evidence of drug-caused changes in the brain that make patients more psychotic and more emotionally dysfunctional.[50]

Nor does the pathology induced by the drugs end there. Still other MRI studies have found that neuroleptic use is associated with shrinkage of the frontal and temporal lobes, and that the risk of frontal atrophy increases 6.5 percent for each ten grams of neuroleptics taken. Such neurodegenerative processes have also been linked to neuroleptic dosage, with higher dosages accelerating the brain damage.[51]

A final risk with neuroleptics is early death. Although this was never the subject of much study, patients kept on the drugs naturally suffered from poor health. The weight gain associated with neuroleptic use increased their risk of diabetes and cardiovascular disease. Nearly 90 percent of medicated schizophrenics also smoked, and one thought was that they did so because the nicotine both lowered neuroleptic blood levels and increased dopamine activity in the brain, and thus ameliorated some of the drugs' nasty effects. But the constant smoking increased their risk of dying from cancer, respiratory illness, and heart disease. Researchers also found a higher mortality rate in patients with tardive dyskinesia and in those patients put on two or more neuroleptics concurrently, which was a common practice.[52]

Thus, over time, neuroleptics affected people in a fairly consistent way. The immediate effects of neuroleptics, which partially shut down dopamine pathways in the brain, were a blunting of emotions and retarded movement. Within a short period, the brain compensated for this drug-induced pathology by becoming supersensitive to dopamine, which made the person more biologically vulnerable to psychosis and prone to worse relapses. With potent drugs like Prolixin and Haldol, an inner anxiety and physical jitteriness frequently tormented the patients, which at times spurred assaultive, criminal behavior. Those who successfully stabilized on neuroleptics became ever more socially withdrawn and isolated. Finally, patients were likely to suffer brain damage of various sorts—the basal ganglia might grow in volume while the frontal lobes shrank—and this would, with some regularity, lead to a more global dysfunction, visible in the muscle spasms common to tardive dyskinesia. The person's physical health would likely decline as well, increasing the likelihood of early death.

By any standard, "medicated schizophrenia" was not a kind fate.

8

THE STORY WE TOLD OURSELVES

How then are we to help "schizophrenics?" The answer is simple: Stop the lies!

—John Modrow[1]

THE STORY OF neuroleptics as drugs that induced a brain pathology, similar in kind to encephalitis lethargica and Parkinson's disease, is one that can easily be dug out from the medical literature. It's all there—the early comparisons to those two diseases, the biological explanation of how neuroleptics sharply impaired dopamine transmission, the importance of that dopamine activity to normal brain function, and the array of deficits, both short-term and long-term, produced by that drug-caused pathological process. Yet that story is not one that American psychiatry, once it had embraced neuroleptics in the early 1960s as safe and effective, was poised to tell, either to itself or to the American public. The country had put its faith in the drugs, and doctors were understandably intent on perceiving the drugs as effective, and at least somewhat safe. But maintaining that belief aloft required mental juggling of the most agile sort, and more than a little talent for self-delusion, poor science, and—ultimately—outright deceit.

A Tale of Biology

Mad-doctors, of course, have always constructed "scientific" explanations for why their treatments worked. Benjamin Rush drained the blood from his patients and reasoned that madness was caused by too much blood flow to the brain. Henry Cotton removed his patients' teeth and other body parts and argued that it cleansed them of madness-causing bacteria. Manfred Sakel stumbled on insulin coma as a treatment and concluded that the treatment miraculously killed the "diseased" brain cells responsible for psychosis. Lobotomy, Egas Moniz said, destroyed nerve fibers where obsessive, paranoid thoughts were stored. Once it was learned that neuroleptics blocked dopamine receptors, psychiatrists reasoned that schizophrenics likely suffered from overactive dopamine systems. The treatment begat the theory of illness, and not vice versa.

As a result of this hypothesis, by the early 1970s patients and their families were regularly hearing this spiel: "I would explain that mental illness is caused by a chemical imbalance in the brain," recalled Susan Kemker, a staff psychiatrist at North Central Bronx Hospital in New York. "Mental illness resembles diabetes, which involves a chemical imbalance in the body, I would explain. The patient's psychiatric disorder is chronic, I would say, and requires medication every day for the rest of the person's life. I would then assure the patient that if he took the medication, he would probably live a more normal life."[2]

Although neuroleptics clearly reduced dopamine activity in the brain to a pathological level, there was still the possibility that schizophrenics started out with hyperactive dopamine systems. Dopamine transmission in the brain works in this manner: A "presynaptic" neuron releases the neurotransmitter into the synaptic cleft (the space between neurons), and then the neurotransmitter binds with receptors on a "postsynaptic" neuron. The dopamine hypothesis suggested that either the presynaptic neurons were releasing too much of the neurotransmitter or else the postsynaptic neurons had too many receptors and thus were "hypersensitive" to dopamine.

To explore the first possibility, investigators measured levels of dopamine metabolites (or breakdown products) in their patients' blood, urine, and cerebrospinal fluid. (Measuring the levels of metabolites provides an indirect gauge of dopamine release in the brain.) One of the first such studies was done in 1974 by Malcolm Bowers, at Yale. He determined that levels of dopamine metabolites in unmedicated schizophrenics were quite normal. "Our findings," he wrote, "do not furnish neurochemical evidence for an overarousal in these patients emanating from a midbrain dopamine system." However, his published results did show one other startling truth: Dopamine turnover markedly increased *after* people were medicated. This was evidence, in essence, of a "normal" brain trying desperately to cope with the drug's blocking of its dopamine signals.[3]

Others soon reported similar findings. In 1975, Robert Post at the NIMH found no evidence of elevated dopamine levels in twenty nonmedicated schizophrenia patients compared to healthy controls. Three different research teams determined in autopsy studies that drug-free schizophrenics apparently had normal dopamine levels. Meanwhile, pharmacologists at St. Louis University School of Medicine and elsewhere fleshed out the pathology in dopamine transmission caused by the drugs. In response to the dopamine blockade, presynaptic dopaminergic neurons apparently went into hyper gear for about three weeks, pumping out more dopamine than normal. Then the brain cells, as if they were burning out, gradually slowed down to the point where they were releasing less dopamine than normal. Some dopaminergic cells turned quiescent, and others began firing in irregular patterns.[4]

There was one other unsettling twist to the dopamine story: A number of research teams, including one at the NIMH, determined that dopamine turnover in some unmedicated chronic schizophrenics was abnormally *low,* which spurred some to characterize schizophrenia as a dopamine-deficiency disease. If so, then neuroleptics would exacerbate this problem. All of this led UCLA neuroscientist John Haracz to gently conclude in 1982: "Direct support [for the dopamine hypothesis] is either uncompelling or has not been widely replicated."[5]

Having failed to find that schizophrenics had abnormally high levels of dopamine, researchers turned their attention to whether their postsynaptic neurons had too many dopamine receptors. At first blush, researchers found just that. In 1978, Philip Seeman and Tyrone Lee at the University of Toronto reported in *Nature* that at autopsy, the brains of schizophrenics had 50 percent or more dopamine receptors than healthy controls. But the patients studied had been on neuroleptics, and, as Seeman and Lee readily acknowledged, it was possible that the neuroleptics had caused the abnormality. Animal studies and other postmortem studies soon revealed that was indeed the case. Investigators in the United States, England, and Germany all determined that taking neuroleptics led to an increase in brain dopamine receptors, and they found little evidence of higher-than-normal receptor levels prior to drug use. "From our data," German investigators wrote in 1989, "we conclude that changes in [receptor density] values in schizophrenics are entirely iatrogenic [drug caused]."[6]

Fifteen years of investigation into dopamine function in schizophrenics had produced a rather disturbing truth. Researchers had speculated that schizophrenics naturally suffered from overactive dopamine systems but had found that this wasn't so. As John Kane, a well-known researcher at Long Island Jewish Medical Center in New York, confessed in 1994, "a simple dopaminergic excess model of schizophrenia is no longer credible." He noted that even Arvid "Carlsson, who first advanced the hypothesis, [has] concluded that there is 'no good evidence for any perturbation of the dopamine function in schizophrenia.' "[7] Yet investigators had found that the drugs profoundly hindered dopamine function and also caused a pathological increase in dopamine receptors in the brain, the very abnormality hypothesized to cause schizophrenia in the first place. In a sense, the drugs were agents that turned a normal brain into a schizophrenic one.

But that story was never told to the public. The public had been sold on a medical paradigm of a different sort, and on August 18, 1996, a consortium of pharmaceutical companies placed an ad in the *New York Times* assuring the public that scientific studies had found that neuroleptics worked just as promised:

> Scientists now know that the causes of schizophrenia and psychosis are often rooted in powerful chemicals in the brain called neurotransmitters. One of these neurotransmitters is dopamine. Schizophrenia and psychosis can result when the brain has abnormal dopamine levels. Because of recent advances, drugs that are able to alter dopamine levels free many patients from the terrible effects of mental illness.[8]

A scientific hypothesis, genuine in kind, had finally given way to a bald-faced lie.

They Do Prevent Relapse, Don't They?

The dopamine hypothesis was one part of the science tale constructed, from the 1960s forward, that maintained the image of neuroleptics as helpful medications. A second part of the story was that the drugs had been repeatedly proven to be effective in two ways: They knocked down acute episodes of psychosis and greatly lowered the risk that patients would relapse. In his 1983 book *Surviving Schizophrenia,* E. Fuller Torrey explained to families: "The fact that antipsychotic drugs work is now well established. They reduce symptoms of the disease, shorten the stay in the hospital, and reduce the chances of rehospitalization dramatically."[9]

Yet, as even mainstream psychiatry began to obliquely confess in the 1990s, this claim of efficacy had been built upon a scientific house of cards.

When a new medical treatment comes along, the usual thing that researchers do is compare it to existing remedies (as well as to placebo). Before neuroleptics arrived, sedatives of various kinds had long been used in asylum settings to curb acute psychotic episodes and were regularly said to be fairly effective. In the 1800s and early 1900s, numerous articles appeared in medical journals touting the benefits of opium, barbiturates, and bromides. One would expect, then, that by the 1980s there would be numerous studies in the medical literature documenting the superiority of neuroleptics. Yet in 1989, when Paul Keck and other Harvard Medical School physicians scoured the literature for well-designed studies that compared the efficacy of neuroleptics to sedatives over a controlled period of time, they could find only *two*. And in those studies, "both treatments produced similar symptomatic improvement in the first days, and perhaps weeks, of treatment."[10] Their report was so unsettling to accepted wisdom that one physician wrote in stunned response: "Has our clinical judgment about the efficacy of antipsychotics been a fixed, encapsulated, delusional perception . . . Are we back to square one in antipsychotic psychopharmacology?"[11] Forty years after neuroleptics were introduced, and still there was no convincing proof that the drugs were any better at knocking down psychosis than old-fashioned opium powder.*

At first glance, the second part of the efficacy question—do neuroleptics prevent relapse?—seems to be a very confused issue. On the one hand, the studies by Bockoven, Carpenter, Rappaport, and Mosher indicate that the use of neuroleptics increases the risk of relapse. Yet at the same time, there are scores of studies in the medical literature that have seemingly made just the opposite conclusion. In 1995, Patricia Gilbert and her colleagues at the University of California at San Diego reviewed sixty-six relapse studies, involving 4,365 patients, and summed up the collective evidence: Fifty-three percent of patients withdrawn from neuroleptics relapsed within ten months, versus 16 percent of those maintained on the drugs. "The efficacy of these medications in reducing the risk of psychotic relapse has been well documented," they wrote.[12]

There is an answer to this puzzle, and it is a revealing one. Bockoven found low relapse rates in patients who had *never been*

---

*A review of seven other studies comparing neuroleptics to benzodiazepines, which are minor tranquilizers, adds to the questions raised by Keck's study. In four trials there was no difference between neuroleptics and the minor tranquilizers; twice the benzodiazepines came out slightly on top; and the neuroleptics did once. See Owen Wolkowitz, "Benzodiazepines in the Treatment of Schizophrenia: A Review and Reappraisal," *American Journal of Psychiatry* 148 (1991), specifically the table on p. 716 for comparative results in six trials; and William Carpenter, "Diazepam Treatment of Early Signs of Exacerbation in Schizophrenia," *American Journal of Psychiatry* 156 (1999):299–303.

exposed to neuroleptics. In a similar vein, the studies by Rappaport, Mosher, and Carpenter involved patients who, at the start of the experiment, were not on neuroleptics but were then treated either with placebo or a neuroleptic. And in those studies, relapse rates were lower for the placebo group. In contrast, the sixty-six studies reviewed by Gilbert were *drug-withdrawal* studies. In those trials, patients stabilized on neuroleptics would be divided into two cohorts: One would keep on taking the drugs and the other would not, and the studies reliably found that people withdrawn from their neuroleptics were more likely to become sick again. Thus, together these studies suggest that relapse rates fell into three groups: lowest for those not placed on neuroleptics in the first place, higher for those who took the drugs continuously, and highest of all for those withdrawn from the drugs.

However, there's still more to be added to this relapse picture. The studies reviewed by Gilbert were designed in ways that grossly exaggerated the difference in relapse rates between drug-maintained and drug-withdrawn patients. First, in two-thirds of the studies, the patients were *abruptly* withdrawn from neuroleptics, and abrupt withdrawal—as opposed to gradual withdrawal—dramatically increased the risk that patients would become sick again. In response to Gilbert's report, Ross Baldessarini of Harvard Medical School reanalyzed the same sixty-six studies, only he divided the drug-withdrawn cohort into "abrupt-withdrawal" and "gradual-withdrawal" groups. He found that the relapse rate in the abruptly withdrawn group was *three times higher* than in the gradual group. In other words, the high 53-percent relapse rate reported by Gilbert for drug-withdrawn patients was, in large part, created by the design of the sixty-six studies. Indeed, in a further review of the relapse literature, Baldessarini and his Harvard colleagues found that fewer than 35 percent of schizophrenia patients gradually withdrawn from their drugs relapsed within six months and that those who reached this six-month point without becoming sick again had a good chance of remaining well indefinitely. "The later risk of relapsing [after six months] was remarkably limited," the Harvard researchers concluded, and they also provided a biological explanation for why this might be so. After the drugs leave the system, they noted, D_2 receptor densities in the brain may revert back to more normal levels, and once this happens, the risk of relapse decreases, returning to a level that "may more nearly reflect the natural history of untreated schizophrenia."[13]

The second flaw in the sixty-six relapse studies was that the low relapse rate for drug-maintained patients—16 percent over a one-year period—was also an artifact of trial design. In the real world, up to 30 percent of hospitalized patients don't respond to neuroleptics. Among those who do and are discharged, more than one-third relapse within the next twelve months and need to be rehospitalized, even though they reliably take their medications. Thus, fewer than 50 percent of people who suffer a schizophrenic break respond to standard neuroleptics and remain relapse-free for as long as a year after discharge. But the relapse studies, to a large degree, were conducted in this select cohort of good responders. It was this group of patients that would be divided into drug-maintained and drug-withdrawn cohorts, and naturally relapse rates for those who stayed on neuroleptics could be expected to be low. In 1998, Gerard Hogarty at the University of Pittsburgh pointed out just how misleading the drug-maintained relapse rates were: "A reappraisal of the literature suggests a 1-year, post-hospital, relapse rate of 40 percent on medication, and a substantially higher rate among patients who live in stressful environments, rather than earlier estimates of 16 percent."[14]

In sum, the sixty-six relapse studies were biased in ways that provided a totally false picture of the merits of neuroleptics. The studies only compared results for drug-treated patients (as opposed to patients never put on neuroleptics), and even within this model of care, the studies painted a false picture. The relapse rate for the drug-withdrawn group was artificially raised by taking patients abruptly off their medications, while the relapse rate for the drug-maintained group was artificially lowered by selecting patients who had already shown that they could tolerate the drugs fairly well. The utter irrelevance of the studies to real-world care shows up dramatically in rehospitalization rates. By one estimate, more than 80 percent of the 257,446 schizophrenia patients discharged from hospitals in 1986 had to be rehospitalized within two years, a rehospitalization rate much higher than for "never-exposed" patients, or—as can be seen by the data above—for those gradually withdrawn from neuroleptics.[15] The 16-percent relapse rate touted in the medical journals was a helpful number for the tale of efficacy that needed to be woven in support of neuroleptics, but it was a statistic derived from science of the worst sort, and it totally misled the public about what was really happening to drug-treated patients.

See No Evil

The third cornerstone of the story we told ourselves about neuroleptics was that the drugs were relatively safe. In 1964, the NIMH specifically declared that side effects with the drugs were "mild and infrequent . . . more of a matter of patient comfort than of safety." Torrey, in his 1983 book, even reiterated the point, assuring families that "antipsychotic drugs are among the safest group of drugs known."[16] But keeping this part of the story afloat for nearly forty years proved particularly difficult. It required that the FDA and American psychiatry turn a blind eye for as long as possible to evidence that the drugs frequently caused tardive dyskinesia and, on occasion, a fatal toxic reaction called neuroleptic malignant syndrome.

From the very beginning, there had been reason to suspect that neuroleptics would cause long-term harm. In the 1930s, first-generation phenothiazines had been used in agriculture as insecticides and to kill parasitic worms in swine. That was their preclinical history—as agents toxic to bugs and parasites. French chemists then developed chlorpromazine as an agent that could help numb the nervous system during surgery. And once chlorpromazine was used in mental patients, it was observed to cause symptoms similar to Parkinson's disease and encephalitis lethargica. After Smith Kline's success with chlorpromazine, other pharmaceutical companies brought new and more powerful neuroleptics to market by selecting compounds that reliably induced catalepsy—a lack of motor movement—in animals. The agents were neurotoxic by *design*. Then, in 1959, the first report appeared linking neuroleptics to irreversible motor dysfunction. This side effect was given the name tardive dyskinesia a year later, and over the next decade, nine studies found that it affected more than 10 percent of all patients, with one suggesting that it might afflict 40 percent of those who got the medications on a constant basis.[17]

And yet the mentally ill were not being told of this risk.

The mechanism by which the FDA warns the public about drug-related risks is by requiring pharmaceutical companies to detail it on the drug's label. Even a side effect that occurs in only 1 percent of patients is considered common and must be warned about. By this standard, tardive dyskinesia was a very common disorder, and yet, throughout the 1960s, the FDA did not require drugmakers to warn the public. Their drug labels typically devoted a single sentence to possible permanent neurological side effects, didn't mention tardive dyskinesia by name, and—despite the reports in the literature concluding that it could affect up to 40 percent of patients—dismissed such problems as uncommon. In 1968, an NIMH scientist, George Crane, published a review of tardive dyskinesia in the widely read *American Journal of Psychiatry*, and still the FDA didn't sound the alarm. Finally, in 1972—thirteen years after the first case report of tardive dyskinesia appeared in the literature—the FDA asked the drug companies to update their labels.

Psychiatry as a profession was proving equally reluctant to acknowledge this problem. In the early 1970s, Crane began something of a crusade to bring this problem to the fore. He wrote about tardive dyskinesia on several occasions, and yet each time he did, his colleagues responded by suggesting that he was making a mountain out of a molehill. Tardive dyskinesia, wrote Nathan Kline in 1968, is a "rare side effect" that is "not of great clinical significance." Boston psychiatrist Jonathan Cole called Crane a "Cassandra within psychiatry" who was needlessly "foreseeing doom in many aspects of our current scientific and clinical operations." In 1973, even after the FDA had finally started to stir, Minnesota physician John Curran chastised Crane's alarms as "not only premature but misleading" and said that even if the drugs did cause brain damage, that shouldn't be reason for undue concern: "While it is true that any psychosis can remit spontaneously, I honestly do not see how one can withhold a treatment of proved efficacy for fear of inflicting or aggravating putative brain damage." Others chalked up TD to brain damage from earlier therapies, particularly lobotomy and electroshock, or attributed it to the disease. It all prompted Crane to retort: "The majority of clinicians continue to ignore the existence of this complication . . . the neglect of a serious health problem for so many years has deeper roots than mere ignorance of facts."[18]

The deeper roots were, of course, money.

Pharmaceutical companies had the most obvious reason for protecting the image of neuroleptics as safe. The drugs had turned into cash cows, and drug companies were not just selling them for use in the mentally ill. By 1970, more than 50 percent of mentally retarded children in America were being drugged in this way. So were a similar percentage of the elderly in nursing homes. Juvenile delinquents were given the drugs so regularly they referred to them as "zombie juice." All told, 19 million prescriptions were being written annually.[19] Public attention to the fact that they frequently caused irreversible brain damage threatened to derail this whole gravy train.

Psychiatry's motivation for turning a blind eye to tardive dyskinesia was a bit more complex. Prescribing a medication is *the* ritual that defines modern medicine, and thus psychiatry, eager to see itself as a medical discipline, needed to have at its disposal a "safe and effective" drug for schizophrenia. Psychiatrists also compete with psychologists for patients, and their one competitive advantage is that because they are medical doctors, they can prescribe drugs, whereas psychologists can't. They could hardly lay claim to superior curative prowess if their neuroleptics were not just ineffective but brain damaging. Finally, by the early 1970s, all of psychiatry was in the process of being transformed by the influence of drug money. Pill-oriented shrinks could earn much more than those who relied primarily on psychotherapy (prescribing a pill takes a lot less time than talk therapy); drug-company sales representatives who came to their offices often plied them with little gifts (dinners, tickets to entertainment, and the like); and their trade organization, the APA, had become ever more fiscally dependent on the drug companies. Thirty percent of the APA's annual budget came from drug advertisements in its journals, and it also relied on industry "grants" to fund its educational programs. "We have evolved a somewhat casual and quite cordial relationship with the drug houses, taking their money readily," an APA officer, Fred Gottlieb, confessed a few years later. "We persist in ignoring an inherent conflict of interest."[20]

In short, the interests of the drug companies, psychiatrists, and the APA were all in synch, and paying too much attention to tardive dyskinesia could prick the whole neuroleptic balloon.

As Crane sounded the alarm, he never urged that neuroleptics be withdrawn. He simply wanted the APA to mount a massive educational campaign to inform physicians how best to manage this risk. Prescribing lower doses could greatly lessen the odds that it would develop. Early diagnosis of TD and a proper therapeutic response—withdrawal of the drugs—could also minimize the harm done. But in the absence of such education, physicians were regularly treating tardive dyskinesia by *upping* dosages (this would so clamp down on motor movement that the jerky motions would be somewhat stilled). Here was a clear and pressing medical need, one that could spare hundreds of thousands of Americans from drug-induced brain damage. "Mailing informative material to all physicians is essential," Crane pleaded in 1973.[21] And in response, the APA . . . dawdled. Daniel Freedman, editor of the *Archives of General Psychiatry*, angrily wrote that psychiatrists already had at their disposal "considerable data and guidelines to help determine sound judgments."[22] Year after year passed, and the APA made no effort to educate its members. The tally of Americans afflicted with this often-irreversible brain disorder was climbing at a rate of *more than 250 people per day*, and still the APA did nothing.[23] Finally, in 1979, the APA issued a task-force report on the problem . . . and then it dawdled some more. Another six years went by before it sent out a warning letter to its members, and that mailing campaign was launched only after several highly publicized civil lawsuits found psychiatrists (and their institutions) negligent for failing to warn patients of this risk, with damages in one case topping $3 million. As the APA put it in its warning letter: "We are further concerned about the apparent increase in litigation over tardive dyskinesia."[24] Money, or the fear of losing it, had finally put the APA into an educational mood.

This foot-dragging obviously told of a stunning disregard for the mentally ill. But even more perplexing was that even when educational efforts were mounted, they didn't do much good. After Crane gave a talk at a well-known hospital on the need to prescribe lower dosages, he returned six months later to see whether anything had changed. Nothing had. "The impact of my educational efforts on the prescribing habits of the physician has been nil," he bitterly reported.[25] Even state laws requiring physicians to tell their patients about this risk didn't do the trick. More than twenty-five states passed such legislation in the early 1980s, laws that implicitly condemned American psychiatry for failing to fulfill this duty on its own, yet a national survey soon found that disclosure rates were *lowest* in states where it was mandatory.[26] In 1984, Thomas Gualtieri, a physician at the University of North Carolina, summed up the dismal history: "A review of the history of TD demonstrates nothing as clearly as this disconcerting fact: since 1957, published guidelines, scientific articles, presentations at professional meetings and draconian admonitions in the *Physicians Desk Reference* seem to have had little, if any, effect on actual physician behavior with respect to neuroleptic drugs."[27]

The tragic result of this head-in-the-sand attitude has never been fully added up. Mantosh Dewan, of the State University of New York Health Science Center in Syracuse, estimated that during the 1980s, more than 90,000 Americans developed "irreversible TD each year."[28] And the blind eye toward TD was simply part of a larger blindness by American psychiatry toward all of the neurological problems that could be induced by neuroleptics. Akathisia, akinesia (extreme blunting of emotions), Parkinson's—all of these regularly went undiagnosed. One 1987 study found that akathisia was missed by doctors 75 percent of the time. The decades-long avoidance of a side effect called neuroleptic malignant syndrome, meanwhile, led to thousands dying needlessly. This toxic reaction to neuroleptics, which typically develops within the first two weeks of exposure, was first described by French physicians in 1960. Incidence estimates range from .2 percent to 1.4 percent. Patients break into fevers and often become confused, agitated, and extremely rigid. Death can then come fairly quickly. Yet, in the United States, neuroleptic malignant syndrome was not given much attention until the early 1980s. The cost of this neglect shows up dramatically in associated mortality rates before and after 1980: They dropped from 22 percent to 4 percent once it became a topic of concern. Although no researcher has tallied up the needless death count, rough calculations suggest that from 1960 to 1980 perhaps 100,000 Americans died from neuroleptic malignant syndrome and that 80,000 of those patients would have lived if physicians had been advised to look for it all along.[29]

Only in America

Although neuroleptics became standard treatment in all developed countries, European physicians never embraced, at least not with the same enthusiasm, the notion that the drugs were "like insulin for diabetes." In 1985, French pioneer Pierre Deniker, at a talk in Quebec City, summed up the view from abroad. First, he recalled, he and Delay had coined the term neuroleptics, which "horrified" the Americans, as it described a drug that clamped down, in the manner of a chemical restraint, on the central nervous system. The Americans preferred the much more benign term "tranquilizers." But then the Americans had transformed the drugs' image again, from tranquilizers into "antischizophrenics," and that, Deniker said, was perhaps going "too far." While neuroleptics might diminish certain symptoms of schizophrenia, he said, they did not "pretend" to be a treatment for a known biological illness.[30]

That difference in view had also been accompanied, Deniker noted, by a difference in prescribing practices. From the beginning, the Europeans—seeing the drugs as *neuroleptics*—had prescribed low dosages to minimize the harmful side effects. After their initial trials with chlorpromazine, Deniker and Delay had decided that 100 milligrams daily was the best dose. British psychiatrists tried a higher dose of 300 milligrams but found that it produced too many negative effects. In contrast, the first American investigators to test chlorpromazine quickly pushed dosages much higher, so much so that Baylor University's John Vernon Ross-Wright told colleagues in 1955 that he had successfully given his patients 4,000 milligrams per day. This high dose, he said, "saved time" in getting patients stabilized and discharged from the hospital. Other leading American psychiatrists soon echoed his beliefs. Patients on 5,000 milligrams daily were said to be "functioning perfectly." "When in doubt with Thorazine," one psychiatrist advised, "increase the dose rather than decrease it." In 1960, New York's Nathan Kline summed up his rule of thumb: "Massive doses for fairly prolonged periods are essential for successful treatment."[31]

The next step up this drugging ladder came in the 1960s, when Prolixin (fluphenazine) and Haldol (haloperidol) were brought to the market. These drugs, developed by Squibb and Janssen pharmaceutical companies, were fifty times more potent than chlorpromazine. Squibb's injectable formulation of fluphenazine shut down dopaminergic pathways so quickly that doctors dubbed it "instant Parkinson's." As would be expected, both of these drugs often caused severe side effects, and yet these were the drugs that American psychiatry turned to. By the 1980s, more than 85 percent of schizophrenics in the United States were on the high-potency neuroleptics.

Over this same period, American psychiatrists ratcheted up the dosage as well. Average daily doses doubled from 1973 to 1985. In the mid-1980s, patients were routinely discharged from hospitals on haloperidol or fluphenazine at daily dosages equivalent to 1,500 milligrams of chlorpromazine (five times what British doctors had initially deemed too problematic). Moreover, it was psychiatrists, rather than non-psychiatric doctors, who were the high dosers. In the 1970s, both of these physician groups prescribed neuroleptics in roughly equivalent amounts. But then, over the course of a decade in which the risk of tardive dyskinesia became well known, their prescribing practices diverged. Non-psychiatric doctors turned to lower doses, while psychiatrists *upped* theirs. By 1985, American psychiatrists were prescribing neuroleptics at dosages that were four times higher than those prescribed by non-psychiatrists.[32] Such doses, Deniker said at the conference in Quebec City, "are enormous according to our [European] point of view."[33]

The prescribing habits of American psychiatrists seem bizarre until one remembers the "scientific" story that had been told about neuroleptics. They were antischizophrenic medications that prevented relapse. High doses—as long as they weren't withdrawn—

best achieved that goal. As Torrey assured families in 1983, the more potent drugs were "better." Indeed, investigators at the University of Pittsburgh, studying this issue, concluded that American psychiatrists often adopted such practices to avoid criticism. By prescribing a potent antischizophrenic drug at a high dose, a psychiatrist could be seen by the patient's family as "doing everything possible" to help.[34]

As usual, though, it was the patients who bore the cost of this delusion. The harm from the high doses was documented in study after study. When high-dose regimens were compared to low-dose regimens, high-dose patients were found to suffer more from depression, anxiety, motor retardation, emotional withdrawal, and akathisia. The incidence of dystonia—painful, sustained muscle spasms—soared. Although high doses would forestall relapse, when patients on such regimens finally did relapse, they often became more severely ill. High doses of fluphenazine were tied to an increased risk of suicide. Even moderately high doses of haloperidol were linked to violent behavior. Van Putten determined that patients placed on a daily 20-milligram dose of Haldol, which was a standard dose in the 1980s, regularly suffered "moderate to severe" akathisia and, by the second week, "deteriorated significantly" in terms of their ability to respond emotionally to the world, and to move about it. This dosage of Haldol, Van Putten concluded, was "psychotoxic" for many patients.[35] As for tardive dyskinesia, it became a common problem for American patients, whereas in Europe, Deniker noted, it "is known but does not have the same importance in severity and in quality."[36]

Together, all of these historical pieces add up to a very dark truth. Neuroleptics, peddled to the public as medications that "altered" dopamine levels in ways that "freed patients from the terrible effects of mental illness," actually induced a pathology similar to that caused by encephalitis lethargica. And American psychiatrists, for more than thirty years, prescribed such drugs at virulent doses.

9

SHAME OF A NATION

They called me mad, and I called them mad, and damn them, they outvoted me.

—Nathaniel Lee[1]

THE ALCHEMY THAT transformed neuroleptics into antischizophrenic medications had, in essence, set two "realities" in motion. There was the reality that the patients experienced and the one that we as a society believed in, and they were in dramatic conflict. During the 1970s, the battle over which reality was "true" spilled into the courts and deep into the hallways at the NIMH. Patients demanded the right to forgo "treatment," and at the NIMH, the head of the schizophrenia division, Loren Mosher, put the question of whether patients might do better without neuroleptics under an experimental microscope. These two struggles marked the proverbial fork in the road, as they raised fundamental questions about the values that would, in the future, drive our care of the mentally ill. Would we be willing to listen to the mentally ill and fashion a form of care responsive to their wants and needs? Would we be willing to honestly explore alternatives to drug treatment?

Or would we simply insist that our medications for schizophrenia were good, and leave it at that?

The answers to those questions can be clearly seen today.

Until patients mounted their legal protests in the 1970s, American society had always pretty much taken for granted that it had the right to forcibly treat the mentally ill. There had been a number of legal battles in the 1800s and early 1900s over society's right to commit patients, but once patients were so committed, forced treatment seemed to follow as a matter of course. Mental patients lacked competence to consent, and—or so the argument went—the state had the right, in the absence of such competence, to act as a substitute parent and determine what was best for them. While there was an understandable rationale to that argument—how can a psychotic person evaluate a proposed treatment?—the history of mad medicine also showed that it invited abuse. Asylum patients had been strapped to tranquilizer chairs and bled, forcibly sterilized, and chased down hallways so they could be injected with metrazol or convulsed with a jolt of electricity. Freeman was so nonchalant about the practice of forced lobotomies that, in one of his books, he included a photo of a screaming, naked woman being dragged to the operating table. To patients, such treatment could be seen as an assault on who they were.

The introduction of neuroleptics into asylum medicine made for a new chapter in this long-running battle between doctor and mental patient. Very early on, hospital psychiatrists began describing how patients, much to their displeasure, were hiding pills in their cheeks and spitting them into toilets when they weren't looking. Discharged patients were found to be "unwilling to purchase the drug."[2] Various studies determined that 40 percent, 50 percent, and even 60 percent of patients were trying to avoid treatment in this way, leading one psychiatrist to lament: "Drug defectors constitute a large part of the world's psychiatric population."[3] The problem of patient resistance was so pervasive that in the early 1960s, pharmaceutical companies scrambled to develop drug-delivery methods that could circumvent this resistance. One solution, which several firms came up with, was to replace the pills with liquid formulations that were odorless and colorless, which hospital staff could then secretly mix into the patients' food. Ads that Smith, Kline & French ran in psychiatry journals for liquid Thorazine revealed the medical attitude behind this subterfuge:

> Warning! Mental Patients Are Notorious DRUG EVADERS. Many mental patients cheek or hide their tablets and then dispose of them. Unless this practice is stopped, they deprive themselves of opportunities for improvement or remission . . . deceive their doctors into thinking that their drugs have failed . . . and impose a needless drain on their hospital's finances. When drug evaders jeopardize the effectiveness of your treatment program, SPECIFY LIQUID CONCENTRATE THORAZINE. Liquid concentrate is the practical dosage for any patient who resists the usual forms of oral medication. It can easily be mixed with other liquids or semisolid foods to assure ingestion of the drug.[4]

The long-acting injectables, first introduced in 1963, were similarly hailed as a "major tactical breakthrough" that made forced treatment easier. Ads promised doctors that an injectable "puts control of the schizophrenic in your hands . . . lightens responsibilities of the hospital staff . . . saves time, reduces costs in the hospital, clinic, office."[5] After a single injection, Heinz Lehmann advised, resistant patients became "cooperative enough to take whatever drug and whatever mode of drug administration is chosen for them." In discharged patients, he added, injections could be likened to an intrauterine device for preventing pregnancy. "Once the medication is in, the patient is safe for a certain period of time," he said.[6] The fact that such long-acting drugs caused worse side effects was seen to be of little consequence, a small price for patients to pay in return for increasing the likelihood they would remain "medication compliant."

One patient who was so treated was David Oaks, who today is the editor of *Mind Freedom*, an activist newsletter for ex-patients. In 1975, he suffered a psychotic break while an undergraduate at Harvard University: "I was told that I would have to be on drugs the rest of my life, that it was like insulin for diabetes. I was held down when I tried to reject the drugging, put in solitary confinement and forcibly injected. It galvanized me to fight back against this oppression. This forced drugging is a horrible violation of core American values of freedom."[7]

That argument, that forced treatment violated fundamental American values, was the basis of legal challenges by patients for the right to refuse medication. The "mad" groups saw their struggle in historical terms, as a long-overdue battle for their civil rights. The Insane Liberation Front formed in Portland; the Mental Patients' Liberation Project in New York City; and the Network Against Psychiatric Assault in San Francisco. They held demonstrations, organized human rights conferences, and, starting in 1975, took their fight to state courts. Their lawyers argued that forced drugging, whether achieved by injection or by slipping it into the patients' food, was a form of medical assault and battery, constituting "cruel and unusual punishment" and a violation of their constitutional rights to due process and freedom of speech. The patients' rallying cry was: "We need love and food and understanding, not drugs."[8]

That was not a message that mainstream psychiatry was eager to hear. The patients' political activities and their lawsuits stirred the wrath of psychiatrists to no end. Abram Bennett, who had helped pioneer convulsive therapies in America, told the *San Diego Union* that ex-mental patients, who were rising up against both forced drugging and the use of electroshock, were a "menace to society" and warned that if the public listened to them, "then insanity will rule the nation." Alexander Rogawaski, a professor at the University of Southern California School of Medicine, publicly called them "bastards" and compared the Network Against Psychiatric Assault to "a dog that bites on your heels and hinders you in what is obviously a very important job."[9] Leaders in psychiatry spoke of how any curbing of forced treatment would pave the way for the mentally ill "to rot with their rights on" and that meddling judges could not understand that psychosis is "itself involuntary mind control" that "represents an intrusion on the integrity of a human being." Antipsychotic medications, they told the courts, "liberate the patient from the chains of illness."[10] In ordinary times, psychiatry might have won this battle easily. But this fray erupted at the same time that Soviet dissidents were smuggling out manuscripts describing neuroleptics as the worst sort of torture, which, at the very least, presented America with the ticklish problem of explaining how a helpful medication here was a poison over there.

A Matter of Perspective

The first rumblings that the Soviets were using neuroleptics to punish dissidents surfaced in 1969 and burst into public consciousness a year later. Dissidents would be diagnosed with "sluggish schizophrenia," their reformist ideas seen as evidence of their "delusions" and poor adjustment to Soviet society, and then sent to one of twelve special psychiatric hospitals. Although the Soviet practices were outrageous, the United States had every reason to be queasy about being too quick to throw stones over this issue. At the time, the United States shared with the Soviet Union the dubious distinction of labeling a larger percentage of its population "schizophrenic" than all other developed countries. Nor was the diagnosis of schizophrenia in the United States free from political, racial, or class taint. In 1958, the first African-American to apply for admission to the University of Mississippi, Clennon King, was committed to a state mental hospital—any black man who thought he could get into Ole Miss was obviously out of touch with reality.[11] Moreover, in the early 1970s, U.S. institutions were routinely using neuroleptics to quiet the mentally retarded, the elderly, and even juvenile delinquents—in such instances, the drugs were clearly being used for non-psychiatric purposes. Even so, U.S. politicians rose up to condemn the Soviets, and in 1972, the U.S. Senate formally launched an investigation into the Soviets' "abuse of psychiatry for [purposes of] political repression."

What the senators heard chilled them. One expert witness, Canadian psychiatrist Norman Hirt, told of a mélange of treatments used to torment the dissidents. Wet packs, insulin coma, metrazol—all familiar to students of American psychiatry—were three such methods. "The fearfulness of these experiences cannot be described adequately by any words," Hirt said. However, written appeals from Soviet dissidents, which had been smuggled out and given to the Senate, described neuroleptics as the worst torture of all. A person who is given aminazine (a neuroleptic similar to Thorazine), wrote Vassily Chernishov,

> loses his individuality, his mind is dulled, his emotions destroyed, his memory lost . . . as a result of the treatment, all the subtle distinctiveness of a person is wiped away. It is death for creativeness. Those who take aminazine cannot even read after taking it. Intellectually they become more and more uncouth and primitive. Although I am afraid of death, let them shoot me rather than this. How loathsome, how sickening is the very thought that they will defile and crush my soul!

Comparisons were drawn between such forced drug treatment and the medical experiments of Nazi doctor Josef Mengele, all of which led Florida senator Edward Gurney to conclude: "Most horrifying of all in this psychiatric chamber of horrors were the many accounts of the forcible administration by KGB psychiatrists of chemicals which convert human beings into vegetables."[12]

Over the next few years, Soviet dissidents published further details of this "chamber of horrors." Aminazine and haloperidol were the two neuroleptics most commonly used to torment them. In a samizdat manuscript titled *Punitive Medicine*, dissidents described the incredible pain that haloperidol could inflict:

> The symptoms of extrapyramidal derangement brought on by haloperidol include muscular rigidity, paucity and slowness of body movement, physical restlessness, and constant desire to change the body's position. In connection with the latter, there is a song among inmates of special psychiatric hospitals which begins with the words, "You can't sit, you can't lie, you can't walk" . . . many complain of unimaginable anxiety, groundless fear, sleeplessness.[13]

Doctors used neuroleptics, the Soviet dissidents stated, "to inflict suffering on them and thus obtain their complete subjugation. Some political prisoners do recant their beliefs, acknowledge that they are mentally ill, and promise not to repeat their 'crimes' in return for an end to this treatment."[14] American psychiatrists also heard such testimony firsthand. On March 26, 1976, Leonid Plyushch, a thirty-nine-year-old mathematician who had spent several years in the psychoprisons before being freed, spoke at a meeting of the New York Academy of Sciences. That produced this memorable exchange:

> Q. What was the most horrifying aspect of your treatment?
>
> A. I don't know if there are irreversible effects of psychiatric treatment, but all the inmates at Dnepropetrovsk Special Psychiatric Hospital lived in the fear that there would be such effects. They had heard stories of those driven by the treatment into permanent insanity. My treatment, in chronological order, began with haloperidol in big dosages without "correctives" that avoid side effects, essentially as a torture. The purpose was to force the patient to change his convictions. Along with me there were common criminals who simulated [mental] illness to get away from labor camps, but when they saw the side effects—twisted muscles, a disfigured face, a thrust-out tongue—they admitted what they had done and were returned to camp.[15]

Such descriptions stirred newspapers and television networks in the United States to condemn, with great fervor, the Soviets' actions. Not long after Plyushch's testimony, the *New York Times* ran an extensive feature on "Russia's psychiatric jails," in which it likened the administration of neuroleptics to people who weren't ill to "spiritual murder" and "a variation of the gas chamber." Dissidents, the paper explained, had been forcibly injected with Thorazine, "which makes a normal person feel sleepy and groggy, practically turning him into a human vegetable." Neuroleptics were a form of torture that could "break your will."[16]

None of this word choice—torture, Mengele, gas chambers, spiritual murder, human vegetables—could possibly have brought any cheer to Smith, Kline & French, or to other manufacturers of neuroleptics. And with the dissidents' words as a foil, U.S. mental patients were able to make powerful cases, in legal challenges filed in Massachusetts, California, New Jersey, Ohio, and elsewhere, that forced neuroleptic treatment was a violation of their constitutional rights. Some of the details that spilled out during those trials were disturbing in the extreme. Judges heard psychiatrists testify that it was best if mental patients were not told about the drugs' side effects, of how patients would be held down by "goon squads" and given an injection in their buttocks, and of hospitals covering up the fact that many of their mental patients suffered from tardive dyskinesia. In New Jersey, John Rennie, a former aircraft pilot who was said to be highly intelligent, was beaten with sticks by aides at Ancora Psychiatric Hospital when he wouldn't take his drugs. The behavior that had landed him there had an obvious political edge as well—he'd threatened to kill President Gerald Ford. At Fairview State Hospital in Pennsylvania, physicians "would enter the ward with a tray of hypodermic needles filled with Prolixin, line up the whole ward or part of the ward, and administer the drug"—care that was reminiscent of the mass shocking of asylum patients.[17] Yet while the newspaper reports condemned the general mistreatment of the mental patients, the drugs—in this context of American medicine, as opposed to the Soviet Union's abuse of its dissidents—were usually presented as helpful medications. They were, the *New York Times* said in its report on Rennie's lawsuit, "widely acknowledged to be effective."[18]

This reporting accurately reflected how the legal struggle played out in court. Judge Joseph Tauro in Boston handed down the groundbreaking ruling on October 29, 1979: "Whatever powers the constitution has granted our government, involuntary mind control is not one of them, absent extraordinary circumstances. The fact that mind control takes place in a mental institution in the form of *medically sound treatment* of mental disease is not, in itself, an extraordinary circumstance warranting an unsanctioned intrusion on the integrity of a human being" (italics added).[19] Judge Tauro had found a way to simultaneously condemn and embrace American practices. Forced treatment was a violation of the patient's constitutional rights, but "mind control" with neuroleptics was a "form of medically sound treatment of mental disease." The image of neuroleptics as good medicine for the mentally ill had been maintained, and in that sense, the patients' victory turned out to be hollow in the extreme. In the wake of the legal rulings, hospitals could still apply to a court to *sanction* forced treatment of drug-resisting patients (it became a due process issue), and as researchers soon reported, the courts almost inevitably granted their approval. "Refusing patients," noted Paul Appelbaum, a psychiatrist at the University of Massachusetts Medical School, "appear almost always to receive treatment in the end."[20] Moreover, since the drugs were still seen as efficacious, society had little reason to develop alternative forms of nondrug care and could even feel justified in requiring patients living in the community, but in need of shelter and food, to take neuroleptics as a condition of receiving such social support. "I spent a lot of years in community mental health," said John Bola, now an assistant professor of social work at the University of Southern California, "and the clients, in order to stay in the residences, would have to agree to take medication. Even when they were having severe reactions to the medication, staff would sometimes threaten to kick them out of the facility unless they took the drugs."[21]

All too often, this resulted in drug-resistant patients finding themselves with nowhere to turn, and on the run. Such was the case for Susan Fuchs. Raised by a loving Brooklyn family, she'd been a bright child and had earned a degree in mathematics from State University of New York at Binghamton. After graduating, however, she found herself caught in the throes of mental illness. She needed help desperately, but neuroleptics only deepened her despair, so much so that at one point early in her illness, she leaped into the Hudson River in a suicide attempt. "I am a vegetable on medication," she wrote. "I can't think. I can't read. I can't enjoy anything . . . I can't live without my mind." That day she was rescued by a bystander, but her fate was cast: She was deeply in need of help, and yet the "help" that society was poised to offer were medications she detested. For the next fifteen years, she cycled in and out of New York's chaotic mental-health system, moving endlessly among psychiatric wards, emergency rooms, and homeless shelters, where she was sexually assaulted. Finally, shortly after midnight on July 22, 1999, a woman's screams were heard in Central Park—the last cry of Susan Fuchs for help. Nobody called the police, and the next morning she was found murdered. Her clothes had been torn from her body, and her head had been bashed in with a rock.[22]

The Defeat of Moral Treatment

The other defining political battle that occurred in the 1970s came in the form of an experiment, known as the Soteria Project, led by Loren Mosher. In their protests, ex-patients had declared that they wanted "love and food and understanding, not drugs," and the Soteria Project, in essence, was designed to compare outcomes between the two. And while love and food and understanding proved to be good medicine, the political fate of that experiment ensured that the Soteria Project would be the last of its kind and that no one would dare to investigate this question again.

Mosher, a Harvard-trained physician, was not "against" neuroleptics when he conceived Soteria. He'd prescribed them while an assistant professor at Yale University, where he'd supervised a ward at a psychiatric hospital. But by 1968, the year he was appointed director of the Center for Schizophrenia Studies at the NIMH, he'd become convinced that their benefits were overhyped. In his new position, he also perceived that NIMH research was skewed toward drug studies. There was, he said, a "clubby" relationship between the academics who sat on the NIMH grant-review committees and the pharmaceutical industry.[23] He envisioned Soteria as an experiment to test a simple premise: Would treating acutely psychotic people in a humanistic way, one that emphasized empathy and caring and avoided the use of neuroleptics, be as effective as the drug treatment provided in hospitals?

Mosher's interest in this question was prompted by a conception of schizophrenia at odds with prevailing biological beliefs. He thought that psychosis could arise in response to emotional and inner trauma, and that it could, in its own way, be a coping mechanism. The "schizophrenic" did not necessarily have a broken brain. There was the possibility that people could grapple with their delusions and hallucinations, struggle through a schizophrenic break, and regain their sanity. His was an optimistic vision of the disorder, and he believed that such healing could be fostered by a humane environment. Soteria would provide a homelike shelter for people in crisis, and it would be staffed not by mental-health professionals but simply by people who had an evident empathy for others, along with the social skills to cope with people who could be strange, annoying, and threatening. "I thought that sincere human involvement and understanding were critical to healing interactions," he recalled. "The idea was to treat people as people, as human beings, with dignity and respect." To give his notions a more rigorous test, he designed the experiment so that only young, unmarried acutely ill schizophrenics would be enrolled—a subgroup that was expected to have poor outcomes.

The twelve-room Soteria house, located in a working-class neighborhood of Santa Clara, California, opened in 1971. Care was provided to six "residents" at a time. When they arrived, they presented the usual problems. They told of visions of spiders and bugs coming from the walls, or of being the devil, or of how the CIA was after them. They could be loud, they could be aggressive, and certainly they could act in very crazy ways. One of the first residents was an eighteen-year-old woman so lost to the world that she would urinate on the floor. She had withered to eighty-five pounds, wouldn't bathe or brush her teeth, and would regularly fling her clothes off and jump into the laps of male staff and say, "Let's fuck." However, faced with such behavior, Soteria staff never resorted to wet packs, seclusion rooms, or drugs to maintain order. And over the course of a decade, during which time more than 200 patients were treated at Soteria and at a second house that was opened, Emanon, violent residents caused fewer than ten injuries, nearly all of them minor.

The philosophy at Soteria was that staff, rather than do things "to" the residents, would "be with them." That meant listening to their crazy stories, which often did reveal deeper stories of past trauma—difficult family relationships, abuse, and extreme social failure. Nor did they try to argue the residents out of their irrational beliefs. For instance, when one resident proclaimed that aliens from Venus had selected him for a secret mission and were going to come to a nearby park at daybreak to pick him up, a staff member took him to the park at the appointed time. When the extraterrestrial visitors didn't arrive, the resident simply shrugged and said, "Well, I guess they aren't going to come today after all," and then returned to Soteria House, where he fell sound asleep.

That was a reality check that had helped psychosis loosen its grip.

Beyond that, the Soteria staff let the residents know that they expected them to behave in certain ways. The residents were expected to clean up. They were expected to help with such chores as cooking. They were expected to not be violent toward others. The staff, in essence, was holding up a mirror, much as the York Quakers had done, that reflected to the residents not an image of madness, but one of sanity. Friendships blossomed, residents and staff played cards and games together, and there were no locks on the doors. Other activities included yoga, reading to one another, and massage.

Not too surprisingly, Soteria residents often spoke fondly of this treatment. "I took it as my home," said one, in a film made at the time. "What is best is nobody does therapy," said another. "We ought to have a whole lot of Soterias," said a third. One of the stars of that film was the young woman who, when she'd arrived at Soteria, had regularly invited men to have intercourse with her—she had blossomed into a striking and poised woman, on her way to marrying a local surfer and becoming a mother. When residents recovered to the point they could leave, they were said to have "graduated," and staff and other residents would throw a small party in their honor. The message was unmistakable: They would be *missed.* Schizophrenics! Said one young man on the day of his graduation: "If it wasn't for this place, I don't know where I'd be right now. I'd have to be on the run if it wasn't for Soteria . . . Soteria saved me from a fate worse than death. Food's good too. And there is a whole lot of love generated around this place. More so than any other place I've been."

By 1974, Mosher and his colleagues were ready to begin reporting outcomes data. As they detailed in several published papers, the Soteria patients were faring quite well. At six weeks, psychotic symptoms had abated in the Soteria patients to the same degree as in medicated patients. Even more striking, the Soteria patients were staying well longer. Relapse rates were lower for the Soteria group at both one-year and two-year follow-ups. The Soteria patients were also functioning better socially—better able to hold jobs and attend school.[24]

And that was the beginning of the end for Mosher and his Soteria project.

Even though Mosher was a top gun at NIMH, he'd still needed to obtain funding for Soteria from the grants committee that oversaw NIMH's extramural research program. Known as the Clinical Projects Research Review Committee, it was composed of top academic psychiatrists, and from the beginning, when Mosher had first appeared before them in 1970, they had not been very happy about this experiment. Their resistance was easy to understand: Soteria didn't just question the merits of neuroleptics. It raised the question of whether ordinary people could do more to help crazy people than highly educated psychiatrists. The very hypothesis was offensive. Had anyone but Mosher come forward with this proposal in 1970, the Clinical Projects Committee probably would have nixed it, but with Mosher, the group had been in a difficult political situation. Did it really dare turn down funding for an experiment proposed by the head of schizophrenia studies at the NIMH? The committee approved the project, but it knocked down Mosher's original request for $700,000 over five years to $150,000 over two years.[25]

With that limited funding, Mosher had struggled to get Soteria off the ground. He also had to fight other battles with the review committee, which seemed eager to hamstring the project in whatever way it could. The committee regularly sent auditors to Soteria because it had doubts "about the scientific rigor of the research team." It repeatedly requested that Mosher redesign the experiment in some fashion. In one review, it even complained about how he *talked* about schizophrenia. Mosher and his colleagues, the committee wrote, liked to espouse "slogans" such as psychosis is a "valid experience to be taken seriously." Then, in 1973, it reduced funding for Soteria to $50,000 a year—a sum so small that it seemed certain to provide Soteria with the financial kiss of death.

At that point, Mosher ran an end run around the clinical projects group. He applied for funding from a division of the NIMH that oversaw the delivery of social services to the mentally ill (housing, and so on), and the peer-review committee overseeing grants for that purpose responded enthusiastically. It called Soteria a high-priority investigation, "sophisticated" in its scientific design, and approved a grant of $500,000 for five years for the establishment of a second Soteria house, which Mosher named Emanon.

The battle lines were now clearly joined. Two different review committees—and one was slinging arrows at Mosher as a scientist and the other praising him for running the experiment in a sophisticated manner. The stakes were clearly high. The very credibility of academic psychiatry, along with its medical model for treating schizophrenia, was on the line. Patients were publicly complaining that neuroleptics were a form of torture, and now here was the physician who was the nation's top official on schizophrenia, and also the editor-in-chief of *Schizophrenia Bulletin* (a prominent medical journal), running an experiment that could provide scientific legitimacy for their complaints. Even the NIMH grants committee that had approved funding for Emanon had acknowledged as much: Soteria, it wrote, was an attempt at a "solution" that could humanize the "schizophrenic experience . . . the need for [an] alternative and successful treatment of schizophrenia is great."

And so when Mosher began to report good outcomes, the clinical projects committee struck back in the only way it could. "The credibility of the pilot study data is very low," the review committee snapped. The study, it said, had "serious flaws." Evidence of superior outcomes for the Soteria patients was "not compelling." Then the committee hit Mosher with the lowest blow of all: It would approve further funding only if he was replaced by another investigator, who could then work with the committee to redesign the experiment. "The message was clear," Mosher says, still bitter twenty-five years later. "If we were getting outcomes this good, then I must not be an honest scientist."

The irony was that Mosher was not even doing the outcomes assessment. Outcomes data—for both Soteria and a comparison group of patients treated conventionally in a hospital setting with neuroleptics—were being gathered by an independent group of reviewers. Mosher well knew that experimenter bias regularly plagued drug studies, and so he'd turned to independent reviewers to rid the Soteria experiment of that problem. Even so, the project was taken away from him. A new principal investigator was recruited to lead the Soteria experiment; it limped along for a few more years, and then in 1977, the clinical projects committee voted to shut down the project. It did so even while making a final grudging admission: "This project has probably demonstrated that a flexible, community based, non-drug residential psychosocial program manned by non-professional staff can do as well as a more conventional community mental health program."*

Soteria soon disappeared into the APA's official dustbin, an experiment that American psychiatry was grateful to forget. However, it did inspire further investigations in several European countries. Swiss physicians replicated the experiment and determined that Soteria care produced favorable outcomes in about two-thirds of patients. "Surprisingly," the Swiss researchers wrote in 1992, "patients who received no or very low-dosage medication demonstrated significantly better results."[26] Ten or so Soteria homes have sprung up in Sweden, and in both Sweden and Finland, researchers have reported good outcomes with psychosocial programs that involve minimal or no use of neuroleptics.

As for Mosher, his career sank along with the Soteria project. He became branded as anti-science, someone standing in the way of the progress of biological psychiatry, and by 1980 he had been pushed out of NIMH. Others who dared question the merits of neuroleptics in the 1970s also quickly discovered that it was a singularly unrewarding pursuit. Maurice Rappaport, who'd found in

---

*After the committee's 1977 decision, Soteria researchers reapplied in 1978 for NIMH funds, and the project was revived for a few more years. But the 1977 decision effectively marked the end of Soteria as an experiment that might threaten mainstream psychiatry. The project had been so hobbled that no data gathered in the post-1975 years were published over the next twenty years, and much of the data—because of a lack of funding—wasn't even analyzed. The blackout kept from the public this finding: John Bola, an assistant professor at the University of Southern California, recently reanalyzed all of the Soteria data, including the post-1975 data, and he determined that the superior outcomes for the Soteria group, compared to those treated conventionally with neuroleptics, "is startling. It looks better than what Mosher published." Only 31 percent of Soteria patients who continued to avoid neuroleptics after leaving Soteria relapsed during a two-year follow-up period, compared to 68 percent of those treated conventionally with neuroleptics. (Relapse rates, however, were high for those Soteria patients who, after they left Soteria House, were placed on neuroleptics by their doctors.)

his study that schizophrenics treated without neuroleptics fared better, was able to get his results published only in a relatively obscure journal, *International Pharmacopsychiatry,* and then he let the matter drop. Crane, who'd blown the whistle on tardive dyskinesia, only to be denounced as an alarmist, left the NIMH and by 1977 was toiling in the backwaters of academic medicine, a clinical professor of psychiatry at the University of North Dakota School of Medicine. In the 1980s, Maryland psychiatrist Peter Breggin took up the cudgel as psychiatry's most vocal critic, writing of the harm caused by neuroleptics and speaking out on television, and he quickly became a pariah, flogged by his peers as "ignorant," an "outlaw," and a "flat-earther." Even the media piled on, with *Time* magazine comparing Breggin to a "slick lawyer" who has "an answer for every argument," one who advances "extremely dubious propositions like the notion that drugs don't help schizophrenics."[27]

No one could have missed the message. American psychiatry and society had its belief system, and it was not about to suffer the fools who dared to challenge it.

Better Off in Nigeria

Mosher's experiment and the court battles had occurred at a very particular time in American history. The Civil Rights movement, protests against the Vietnam War, and Watergate all made the early 1970s a time when disenfranchised groups had a much greater opportunity than usual to be heard. Ken Kesey's book *One Flew over the Cuckoo's Nest* suggested that even crazy people should be listened to. That was the societal context that made it possible for the clash between the two realities—the one experienced by patients and the one we as a society believed in—to momentarily become a matter of public debate. With the demise of the Soteria Project, however, the debate officially ended. The 1970s passed into the 1980s, and lingering protests by patients over their drugs were dismissed as the rantings of crazy people. As Edward Shorter declared in his 1997 book *A History of Psychiatry,* antipsychotic medications had initiated a "revolution" in psychiatry and made it possible for patients with schizophrenia to "lead relatively normal lives and not be confined to institutions."* That became the agreed-upon history, and not even repeated findings by the World Health Organization that schizophrenics in developed countries fared much worse than schizophrenics in poor countries, where neuroleptics were much less frequently used, disturbed it.

The WHO first launched a study to compare outcomes in different countries in 1969, a research effort that lasted eight years. The results were mind-boggling. At both two-year and five-year follow-ups, patients in three poor countries—India, Nigeria, and Colombia—were doing dramatically better than patients in the United States and four other developed countries. They were much more likely to be fully recovered and faring well in society—"an exceptionally good social outcome characterized these patients," the WHO researchers wrote—and only a small minority had become chronically sick. At five years, about 64 percent of the patients in the poor countries were asymptomatic and functioning well. Another 12 percent were doing okay, neither fully recovered

*This is yet another part of the "story" we have told ourselves about neuroleptics that is easily shown to be false. Neuroleptics were introduced into mental hospitals in 1954–1955. At that time, fiscal concerns were driving states to seek alternatives to hospitalization of the mentally ill. Even so, over the next seven years the number of patients in public mental hospitals declined only slightly, from 559,000 in 1955 to 515,000 in 1962. The real emptying of the state hospitals began in 1965 with the enactment of Medicaid and Medicare laws. Those laws provided federal subsidies for nursing home care but no such subsidy for care in state mental hospitals, and so the states did the obvious economic thing: They began shipping their chronic patients to nursing homes. The number of patients in state hospitals declined by nearly 140,000 patients from 1965 to 1970, while the nursing home census rose accordingly. Then, in 1972, the federal government passed welfare legislation that provided social security income payments to the disabled. That enabled state hospitals to discharge patients to boarding homes and welfare hotels, with the federal government then stuck with picking up the cost of that care. The year after the SSI law went into effect, the population in state mental hospitals dropped 15.4 percent, the largest decrease ever. By 1980, the census in public mental hospitals in the United States had declined to 132,164. Four hundred thousand beds had been eliminated in a short fifteen years, but it was a deinstutionalization process that had been driven by fiscal concerns, and not by the arrival of neuroleptics.

nor chronically ill, and the final 24 percent were still doing poorly. In contrast, only 18 percent of the patients in the rich countries were asymptomatic and doing well, 17 percent were in the so-so category, and nearly 65 percent had poor outcomes.[28] Madness in impoverished countries ran quite a different course than it did in rich countries, so much so that the WHO researchers concluded that living in a developed nation was a "strong predictor" that a schizophrenic patient would never fully recover.[29]

These findings, which were first reported in 1979, naturally stung psychiatrists in the United States and other rich countries. But Western doctors were not used to seeing their medicine produce such embarrassing results, so many just dismissed the WHO studies as flawed. The people being diagnosed as schizophrenic in the poor countries, the argument went, must not have been suffering from that devastating disorder at all but from some milder form of psychosis. With that criticism in mind, the WHO launched a follow-up study. This time it compared two-year outcomes in ten countries, and it focused primarily on first-episode schizophrenics, all diagnosed by the same criteria. The WHO investigators even divided patients into schizophrenia subtypes and compared outcomes in the subgroups. But it didn't matter. No matter how the data were cut and sliced, outcomes in poor countries were much, much better. "The findings of a better outcome of patients in developing countries was confirmed," the WHO investigators wrote in 1992.[30] Even the statistics were much the same the second time around. In the poor countries, nearly two-thirds of schizophrenics had good outcomes. Only slightly more than one-third became chronically ill. In the rich countries, the ratio of good-to-bad outcomes was almost precisely the reverse. Barely more than one-third had good outcomes, and the remaining patients didn't fare so well.

The sharply disparate results presented an obvious conundrum. Why should there be such a stark difference in outcomes from the same disorder? Suffer a schizophrenic break in India, Nigeria, or Columbia, and you had a good chance of recovering. Suffer the same illness in the United States, England, or Denmark, and you were likely to become chronically ill. Why was living in a developed country so toxic? The WHO investigators looked briefly at various possibilities that might explain the difference—family involvement, childhood experiences, and societal attitudes—but couldn't come up with an answer. All they could conclude was that for unknown reasons, schizophrenics in developed countries generally failed to "attain or maintain a complete remission of symptoms."

However, there was in the WHO's own data a variable that explained the difference. But it was one so threatening to Western medicine that it went unexplored.

The notion that "cultural" factors might be the reason for the difference has an obvious flaw. The poor countries in the WHO studies—India, Nigeria, and Colombia—are not at all culturally similar. They are countries with different religions, different folk beliefs, different ethnic groups, different customs, different family structures. They are wildly disparate cultures. In a similar vein, the developed countries in the study—the United States, England, Denmark, Ireland, Russia, Czechoslovakia, and Japan—do not share a common culture or ethnic makeup. The obvious place to look for a distinguishing variable, then, is in the *medical* care that was provided. And here there was a clear difference. Doctors in the poor countries generally did not keep their mad patients on neuroleptics, while doctors in the rich countries did. In the poor countries, only 16 percent of the patients were maintained on neuroleptics. In rich countries, 61 percent of the patients were kept on such drugs.

That is a statistically powerful correlation between drug use and outcomes. Certainly if the correlation had gone the other way, with routine drug use associated with much better outcomes, Western psychiatry would have taken a bow and given credit to its scientific potions. American psychiatry, after all, had made continuous medication the cornerstone of its care. Yet, in the WHO studies, that was the model of care that produced the *worst* outcomes. Indeed, the country with arguably the poorest outcomes of all was the Soviet Union, and it was also the country that led all others in keeping patients continually on neuroleptics. Eighty-eight percent of Soviet patients were maintained on the drugs, and yet fewer than 20 percent were doing well at the end of two years.[31]

Even before the 1992 WHO report, American researchers had reason to think that there would be such a correlation. In 1987,

TABLE 9.1
Schizophrenia Outcomes:
Developing vs. Developed Countries

	Developing Countries	Developed Countries
Drug Use		
On antipsychotic medication 76% to 100% of follow-up period	15.9%	61%
Best Possible Outcomes		
Remitting course with full remission	62.7%	36.9%
In complete remission 76% to 100% of follow-up period	38.3%	23.3%
Unimpaired	42.9%	31.6%
Worst Possible Outcomes		
Continuous episodes without complete remission	21.6%	38.3%
In psychotic episodes for 76% to 100% of follow-up period	15.1%	20.2%
Impaired social functioning throughout follow-up period	15.7%	41.6%

SOURCE: *Psychological Medicine*, supplement 20 (1992)

Courtenay Harding, a psychologist at the University of Colorado, reported on the long-term outcomes of eighty-two chronic schizophrenics discharged from Vermont State Hospital in the late 1950s. She had found that one-third of this cohort had recovered completely. And as she made clear in subsequent publications, the patients in this best-outcomes group shared one common factor: They *all* had successfully weaned themselves from neuroleptics. Hers was the best, most ambitious long-term study that had been conducted in the United States in recent times. The notion that schizophrenics needed to stay on medication all their lives, she'd concluded, was a "myth."[32]

The correlation between poor outcomes and neuroleptics also clearly fit with all that was known about the biological effects of the drugs. They induced a pathology in dopamine transmission akin to that caused by Parkinson's disease and encephalitis lethargica. They destabilized dopaminergic systems in ways that made patients more vulnerable to relapse. They caused tardive dyskinesia, an often irreversible form of brain damage, in a high percentage of patients. How could such drugs, when prescribed as long-term, maintenance medications, possibly help mentally fragile people function well in society and fully recover from their descent into psychosis? "You are taking people who are already broken—and by that I mean traumatized, broken by life—and then you are breaking them completely," said David Cohen, a professor of social work at Florida International University.[33]

The WHO studies, however, did more than just challenge American psychiatry to rethink its devotion to neuroleptics. The studies challenged American psychiatry to rethink its whole conception of the disorder. The studies had proven that recovery from schizophrenia was not just possible, but *common*—at least in countries where patients were not continually kept on antipsychotic medications. The WHO studies had demonstrated that the American belief that schizophrenics necessarily suffered from a biological brain disorder, and thus needed to be on drugs for life, wasn't true. Here was a chance for American psychiatry to learn from success in other countries and, in so doing, to readjust its message to people who had the misfortune to suffer a schizophrenic break. Recovery was possible. That was a message that would provide patients with the most therapeutic agent of all: *hope.* They did not need to consign themselves to a future dimmed by mind-numbing medications. And with that conception of the disorder in mind, medical care of the severely mentally ill would presumably focus on helping them live medication-free lives. Either they would never be exposed to neuroleptics in the first place, or if they were, they would be encouraged to gradually withdraw from the drugs. Freedom from neuroleptics would become the desired therapeutic goal.

But, of course, that never happened. American psychiatry, ever so wed to the story of antipsychotic medications, a bond made strong by pharmaceutical money, simply ignored the WHO studies and didn't dig too deeply into Harding's, either. Schizophrenics suffered from a biological brain disorder, antipsychotic medications prevented relapse, and that was that. The tale that had been crafted for the American public was not about to be disturbed. Indeed, a few years after the WHO reported its results, an NIMH-funded study determined that care in the United States was proceeding headlong along a path directly opposite to that in the poor countries: In 1998, 92 percent of all schizophrenics in America were being routinely maintained on antipsychotics.[34] Even the thought of getting patients off the drugs had become lost to the medical conversation—evidence, once again, that American psychiatry was being driven by an utterly closed mind.

10

THE NUREMBERG CODE DOESN'T APPLY HERE

Everything they did to me was for the purposes of their research. As my medical record shows, when I went into the hospital I was calm and cooperative. I was just worried and vulnerable. I came out thinking I was crazy, and my parents thinking I was crazy, and my friends thinking I was crazy. My family and I believed that every psychotic feeling and behavior was natural to me, rather than caused by their experiment.

—Shalmah Prince[1]

THE RECORD OF care provided to the severely mentally ill in America from the early 1950s to the early 1990s, the period when standard neuroleptics were the drugs of choice, is, by even the most charitable standards, a disturbing one. Neuroleptics were not just therapeutically neutral, but clearly harmful over the long term. In the United States (as opposed to in Europe), the harm caused by the drugs was further exacerbated by bad medical practices of various sorts: poor diagnostics, a failure to warn patients about the risk of tardive dyskinesia, and the prescribing of neuroleptics in very high doses. As Harvard Medical School psychiatrist Joseph Lipinski said in 1985, speaking of patients misdiagnosed as schizophrenic who were then put on neuroleptics: "The human wreckage is outrageous."[2] Moreover, there was one other troubling aspect to this medical story, and it too was uniquely American.

This element of bad medicine involved experiments that were done on the mentally ill. To fully understand it, it is necessary to backtrack briefly to 1947. That year, America prosecuted Nazi doctors at Nuremberg and drew up the code that was supposed to guide human experimentation in the future.

The Nuremberg trial that is most familiar to the American public is the first, which focused on Nazi military leaders and their "crimes against humanity." That trial was jointly prosecuted by the Allied countries. The second "Doctors' Trial," which involved the prosecution of German physicians for the deadly experiments they had conducted during World War II, was solely an American affair. American lawyers alone acted as the prosecutors, and the trial was presided over by American judges. The United States, by conducting it, was presenting itself as the country that would insist that science be carried out in a moral manner, as the leader in establishing the boundaries of ethical research.

The Nazi experiments, although ghastly in the extreme, did have recognizable scientific aims. Many were designed to provide information useful for wartime medicine. For instance, the German military was concerned about the fate of its pilots shot down and forced to parachute into frigid North Atlantic seas. In order to study survival strategies and test survival gear, Nazi physicians forced Jews at the Dachau concentration camp to drink seawater and also immersed them in freezing water until they died. They put Jews, Poles, and Russians into pressure chambers to investigate how rapid changes in altitude might affect pilots bailing out of their planes. At Ravensbrueck, Nazi physicians shot people to test blood-coagulation remedies. They also deliberately injured prisoners and then exposed their wounds to bacteria—an experiment designed to test treatments for infections. Furthermore, as historian Robert Proctor has written, the Nazi doctors believed there was a moral basis for their experiments. The eugenic philosophy of the day valued the lives of some as more worthy than others. The Jews and prisoners killed or maimed in such experiments were considered inferior beings, and the knowledge to be gained might save the lives of superior Germans. With such a pecking order at work, Nazi physicians could perform such experiments believing they served some ultimate "good."

In August 1947, American judges found fifteen of the twenty-three defendants guilty and sentenced seven to death by hanging. As part of the judgment, two American physicians, Andrew Ivy and psychiatrist Leo Alexander, wrote the ten-point Nuremberg Code for ethical human experimentation. At the heart of the code, which America promoted as a "natural law" that should be respected by all, was the principle that the interests of science should never take precedence over the rights of the human subject. Research subjects were not to be seen as means to a scientific end, and they needed to always give informed consent. As Holocaust survivor Elie Weisel later wrote: "The respect for human rights in human experimentation demands that we see persons as unique, as ends in themselves."[3]

However, the ink on the Nuremberg Code was barely dry when Paul Hoch, director of research at the New York State Psychiatric Institute, began giving LSD and mescaline to schizophrenics in order to investigate the "chemistry" of psychosis.

First Mescaline, Then Lobotomy

Like a number of biologically oriented psychiatrists in postwar America, Hoch had trained in Europe. Born in Budapest, he attended medical school in Göttingen, Germany, and became a German citizen in 1929. After emigrating to the United States in 1933, he landed a job at Manhattan State Hospital and by 1939 was directing its shock therapy unit. He was a strong advocate for the various somatic therapies of the day, and he coauthored a book—with Lothar Kalinowsky, a fellow German immigrant—extolling these treatments. From 1948 to 1955, he directed the Department of Experimental Psychiatry at the New York State Psychiatric Institute, a post that made him a national leader in schizophrenia research.

One of the challenges that Hoch and others faced was developing a model for schizophrenia. The usual practice in medicine is to develop an animal model for the disease to be studied, but Hoch and others reasoned that this was not a feasible approach with schizophrenia. This was a distinctly human condition. In order to investigate the biology of this ailment, it would be necessary to develop methods for "modeling psychosis" in humans. By using drugs to experimentally produce delusions and hallucinations in the mentally ill, Hoch argued, it would be possible to "elucidate the chemical background of these experimental psychoses."[4]

After testing six compounds, Hoch and his colleagues settled on LSD and mescaline as the psychedelic agents of choice. From 1949 to 1952, they gave these drugs to more than sixty mentally ill patients. Both drugs, Hoch reported, "heightened the schizophrenic disorganization of the individual" and thus "are very important in magnifying the schizophrenic structures in schizophrenic patients." In addition, LSD and mescaline could trigger full-blown schizophrenic episodes in "pseudoneurotics" who, prior to the experiment, "did not display many signs of schizophrenic thinking." Such patients, when given LSD and mescaline, Hoch told a packed audience in Detroit at the APA's national convention in 1950, "suffered intensely," underwent a "marked intellectual disorganization," and "were dominated by their hallucinatory and delusional experiences." This type of research, he later wrote, was "helping to establish psychiatry as a solid fact-finding discipline."

At the convention, Hoch also detailed how he'd studied whether electroshock and lobotomy would block drug-induced psychosis. The people in this experiment underwent a grueling series of assaults on their brains. First, they were injected with mescaline to see how they reacted to the drug. Then they were injected a second time with mescaline and hit with electroshock to see if it would knock out their psychosis. It did not. Finally, a number of them were lobotomized (or underwent a surgical variation known as a topectomy), and then injected once more with mescaline. This was done, Hoch said, "to study the clinical structure [of psychosis] before and after psychosurgery." Mescaline did successfully "reactivate the psychosis," but the lobotomized patients no longer responded to the psychosis with the same emotional fervor.

One of the case studies Hoch detailed was that of a thirty-six-year-old man, who, prior to the experiment, had simply complained of constant tension, social inadequacy, and an inability to relax. He was one of the "pseudoneurotics" who, prior to the experiment, did not display many signs of "schizophrenic thinking." But then he was, in essence, sacrificed for purposes of research:

> Under mescaline he complained about a peculiar taste in his mouth, a feeling of cold, and some difficulty in swallowing. He had some visual hallucinations. He saw dragons and tigers coming to eat him and reacted to these hallucinations with marked anxiety. He also had some illusionary distortions of the objects in the room. The emotional changes were apprehension and fear—at times mounting to panic, persecutory misinterpretation of the environment, fear of death, intense irritability, suspiciousness, perplexity, and feelings of depersonalization . . . The mental picture was that of a typical schizophrenic psychosis while the drug influence lasted. This patient [then] received transorbital lobotomy and was again placed under mescaline. Basically the same manifestations were elicited as prior to the operation with the exception that quantitatively the symptoms were not as marked as before.[5]

By this date, Hoch was not unappreciative of the costs associated with lobotomy. Such surgery, he wrote, "damaged" the personality—it made people apathetic, lacking in will, and emotionally shallow.[6] Others had criticized lobotomy as "amputation of the soul." Even so, at the end of his presentation, Hoch's peers rose to congratulate him for his "careful and imaginative work." Nevada psychiatrist Walter Bromberg observed that mescaline appeared to act as a "magnifying glass" for studying schizophrenia. Added Victor Vogel from Kentucky: "The exciting possibilities of research with experimentally produced psychoses are apparent."[7] Nobody stepped up to question the ethics of such experiments. All anyone saw was that scientific knowledge was being pursued, and soon other American scientists were proudly detailing how they too had given psychedelic agents to the mentally ill. Physicians at Wayne State University, for instance, tried making schizophrenics worse through the use of sensory isolation, sleep deprivation, and phencyclidine, a drug that successfully produced in their patients "profoundly disorganized regressive states" that persisted more than a month.[8] As for Hoch, he rose to ever greater prominence in American psychiatry. He became commissioner of New York's mental health department, was elected president of the Society of Biological Psychiatry, and served as editor in chief of the journal *Comprehensive Psychiatry*. At his death in 1964, he was warmly and widely eulogized as the "complete psychiatrist," and his bust was placed in the lobby of the New York State Psychiatric Institute, a bronze plaque hailing him for being a "compassionate physician, inspiring teacher, original researcher [and] dedicated scientist."

The Dopamine Revival

This line of experimentation, while seen as having great promise in the 1950s, almost came to an end in the early 1960s. In 1962, Leo Hollister, a Veterans Administration (VA) psychiatrist in California, delivered a devastating critique of this "popular tool" in psychiatric research, arguing that LSD, mescaline, psilocybin, and other psychedelics didn't produce a model psychosis at all. Such agents primarily provoked visual hallucinations, whereas schizophrenic patients mostly grappled with auditory delusions.[9] The drugs were simply making them suffer in new ways. Even more problematic, the federal government decreed in the early 1960s that LSD and other psychedelic agents were dangerous, making their sale or possession by the general public a criminal act. Once the government had adopted that stance, it became difficult for the NIMH to give researchers money to administer such chemicals to the severely mentally ill.

However, the dopamine hypothesis breathed new life into symptom-exacerbation experiments. The scientific rationale was easy to follow: If psychosis was caused by overactive dopamine systems (which was a fresh hypothesis at that time), then agents that caused brain neurons to release dopamine—amphetamine, methylphenidate, L-dopa—should theoretically make the severely mentally ill worse. The first researcher to test this premise was David Janowsky, a physician at the University of California at San Diego School of Medicine. In 1973, he reported that amphetamine injections could "rapidly intensify psychotic symptoms in patients who are psychotic or 'borderline' psychotic prior to injection." In a subsequent experiment, he even catalogued the relative potency of three dopamine-releasing drugs—d-amphetamine, l-amphetamine, and methylphenidate—in stirring hallucinations and delusions in the mentally ill. Methylphenidate, which caused a doubling in the severity of symptoms, was tops in this regard.[10]

Janowsky's work was seen as providing evidence for the dopamine hypothesis. Amphetamine could admittedly make "normals" psychotic, but he'd shown that it took smaller doses than usual to worsen the mentally ill. He also provided an ethical justification for the studies: "We believe that one to two hours of temporary intensification of psychiatric symptoms, occurring after infusion of a psychostimulant, can be warranted on occasion by the differential diagnostic and psychotherapeutic insights gained during the induced cathartic reaction."

After that, symptom-provocation experiments became an accepted practice in American psychiatry and remained so for the next twenty-five years. Researchers accepted Janowsky's argument that making mental patients sicker for "transient" periods was ethically acceptable, and by the mid-1980s, more than 750 mentally ill people had been in such government-funded studies. Even findings by other investigators that schizophrenics didn't appear to naturally suffer from abnormal dopamine activity didn't quell this research. At the very least, the symptom-exacerbation experiments suggested that dopamine systems were less "stable" in schizophrenics than in others and that acute psychotic episodes were perhaps associated with transient increases in dopamine activity. Researchers used symptom-exacerbation experiments to probe these possibilities, with the thought that understanding the biology of madness in this nuanced way might one day lead to new tools for diagnosing schizophrenia or, at some point, to better drugs.

Patients were recruited into these studies at different points in their illness. In some experiments, people suffering a first bout of psychosis and coming into emergency rooms for help were studied. Physicians at Hillside Hospital in Queens, New York, for instance, gave methylphenidate to seventy first-episode patients, which, they reported, caused 59 percent of them to temporarily become "much worse" or "very much worse." The patients were then placed on neuroleptics, but they took longer than usual to stabilize: Twelve of the seventy were still psychotic a year later. "We were surprised by the length of time required for patients to recover," the New York doctors confessed.[11] Physicians at the University of Cincinnati Medical Center, meanwhile, reported in 1997 that they had given multiple doses of amphetamines to seventeen first-episode patients, including some as young as eighteen years old, who had been newly admitted to the hospital. They did so to see if the patients would get progressively more psychotic with each amphetamine dose, with the thought that this would provide insight into the "sensitization" process that led to "frank psychosis."[12]

Other doctors studied how challenge agents affected hospitalized patients who had recovered somewhat from their acute episodes of psychosis. Could dopamine-releasing agents reawaken the disease? In 1991, doctors at Illinois State Psychiatric Institute injected methylphenidate into twenty patients who'd been in the hospital for two weeks (some of whom had become asymptomatic and were successfully off neuroleptics) and found that it caused "moderate" or "marked deterioration" in most of them. This proved, they concluded, that "methylphenidate can activate otherwise dormant psychotic symptoms."[13] In a similar vein, physicians at the Medical College of Virginia gave amphetamine to nineteen patients who had recovered to the point that they were ready to be discharged; four significantly worsened, and one became wildly psychotic again.[14]

Yet another group studied in this manner were patients who were living in the community. In 1987, physicians at the Bronx Veteran Administration Medical Center abruptly withdrew neuroleptic medication from twenty-eight schizophrenics—including seven who were not considered "currently ill"—and injected them seven days in a row with L-dopa, which had been shown in 1970 to stir hallucinations and other psychotic symptoms. The Bronx doctors wanted to see if those patients who most worsened in response to the L-dopa would then fall into a full-fledged relapse the quickest. All twenty-eight patients eventually descended back into psychosis, including one person who had been stable for fifteen years prior to the experiment.[15]

As researchers reported their results in such journals as *Biological Psychiatry, Archives of General Psychiatry,* and *American Journal of Psychiatry,* they generally didn't describe how individual patients had fared. Instead, they usually reported on the patients in the aggregate—tallying up the percentage of patients who had been made worse. However, in 1987, NIMH scientists broke this mold, detailing how methylphenidate injections had stirred episodes of "frightening intensity" in patients. They wrote of one man:

> Within a few minutes after the [methylphenidate] infusion, Mr. A. experienced nausea and motor agitation. Soon thereafter he began thrashing about uncontrollably and appeared to be very angry, displaying facial grimacing, grunting and shouting. Pulse and blood pressure were significantly elevated . . . Fifteen minutes after the infusion he shouted, "It's coming at me again—like getting out of control—it's stronger than I am." He slammed his fists into the bed and table and implored us not to touch him, warning that he might become assaultive.[16]

Remarkably, even that vivid account of a patient's suffering didn't derail this line of research. Instead, the pace of this experimentation *accelerated* in the 1990s. The invention of new imaging techniques, most notably positron emission tomography (PET), made it possible for researchers to identify the brain regions most active during psychotic episodes, and so they turned to amphetamines and other dopamine-releasing chemicals to provoke this psychosis on cue. In addition, new drugs were coming to market, known as "atypical" antipsychotics, that altered both dopamine and serotonin levels, and this led to speculation that a number of neurotransmitters were involved in mediating psychosis—dopamine, serotonin, glutamate, and norepinephrine. To explore the role of these other neurotransmitters, researchers turned to new chemical agents to exacerbate symptoms in schizophrenics.

At Yale University, for example, doctors injected twelve schizophrenics at a VA hospital with m-chlorophenylpiperazine, a chemical that affected serotonin activity. As they had hypothesized, "characteristic symptoms for each patient worsened."[17] Psychiatrists at both NIMH and the University of Maryland, meanwhile, explored the effects of ketamine—the chemical cousin of the notorious street drug "angel dust"—on schizophrenic symptoms. This drug, which alters glutamate activity in the brain, was found by NIMH scientists to worsen positive symptoms, negative symptoms, and cognitive function and thus appeared to provide a better model of schizophrenia than amphetamines did, as it exacerbated a broader range of symptoms.[18] The Maryland researchers also reported that the worsening of symptoms with ketamine persisted for hours, and at times into the next day. They wrote of one twenty-eight-year-old man:

> On ketamine, he experienced both an increase in disorganized thoughts (neologisms, flight of ideas, loose association), suspiciousness, and paranoid delusions. At 0.1 mg. he became mildly suspicious; at 0.3 mg. he presented moderate thought disorganization and paranoid delusions; and at 0.5 mg. he was floridly delusional, commenting on how he rescued the president of the United States from an assassination attempt.[19]

Symptom-exacerbation experiments—funded by taxpayers and conducted by psychiatrists at some of the leading medical schools in the country—were almost exclusively an American affair. European investigators in the 1970s, 1980s, and 1990s did not publish similar accounts in their medical journals. For the longest while, the experiments were also conducted in relative obscurity, unnoticed by the general public. However, in the mid- to late 1990s, a citizens group called Circare, led by Vera Sharav, a Holocaust survivor whose son had died from a toxic reaction to neuroleptics, and Adil Shamoo, a biology professor at the University of Maryland School of Medicine, whose son is ill with schizophrenia, began shining a public light on this research. "These types of experiments," Sharav protested, "could only be done on the powerless."[20] America's psychiatric researchers were suddenly on the hot seat. Why would anyone volunteer to be in such studies?

The answer the investigators put forth was fascinating, for it required the public to think of schizophrenics in a whole new light.

The reason that schizophrenia patients volunteered for symptom-exacerbation experiments, several researchers said, was that they wanted to make a contribution to science and took satisfaction from doing so. Circare and others who would stop this research, the researchers added, would be denying the mentally ill this opportunity. "In a free country like this," explained David Shore, associate director for clinical research at NIMH, "people have a right to take a risk. They have a right to go through a temporary increase in symptoms if they believe it will be beneficial to understanding the causes of disease. I often say that mental disorders and altruism are not mutually exclusive. It shortchanges the humanity of people who have some of these disorders to say that we are not going to allow them to participate in any studies to get at the underlying causes of the disorder."[21]

The public had a mistaken understanding of the severely mentally ill and their capacity for rational thought, several researchers said. Even people actively psychotic, showing up at an emergency room for help, could retain the presence of mind to give "informed consent" to an experiment designed to exacerbate their symptoms. "Patients who are having psychotic symptoms often can function quite well in many areas of their lives," said Paul Appelbaum, chairman of the psychiatry department at the University of Massachusetts Medical School. "They may have delusions and odd ideas about the CIA investigating their backgrounds, or the FBI trailing them on the street. But that doesn't prevent them from understanding what they need to buy at the supermarket that night to make dinner, or to understand what is being proposed regarding their entering into a research study."[22] Added University of Cincinnati psychiatrist Stephen Strakowski: "If you work with these patients, the vast majority are clearly capable of discussing any research protocol and making a reasonable decision. It is a stigmatizing view that people with mental illness can't make that decision."[23]

There was one flaw with their explanation, however. It was belied by the paper trail for the experiments.

Without Consent

Even though the Nuremberg Code required that all volunteers give "informed consent," American psychiatrists conducting symptom-exacerbation experiments rarely addressed this topic prior to the

1980s. One reason for that, as became clear in 1994, was that they had concluded it was best not to tell their patients that the experiments might make them sicker. This startling confession—that the patients were simply being left in the dark about the nature of the experiments—came from Dr. Michael Davidson, who had led the Bronx Veterans Administration L-dopa studies.

Early in 1994, Sharav obtained the informed consent form for patients in the L-dopa study. In the form, Davidson did not tell his patients that L-dopa, in previous studies, had been shown to provoke a "toxic psychosis." Instead, the consent form stated that the purpose of the experiment was to "measure blood levels of various brain hormones released by [L-dopa]," and that, in this manner, "we may be able to tell if your regular medication is safe for you." As to any risk patients might face, Davidson and his colleagues noted that while L-dopa could cause an increase in blood pressure or an upset stomach, "we do not anticipate any such side effects from the doses we will administer in this study."[24] It was quite evident that the consent form misled patients, and Davidson explained in a letter to Sharav why this was so. Back in 1979, he wrote, when the study had been conceived, the research community believed:

> It would not be advisable to talk to the patients about psychosis or relapse. This explains why language such as "to examine if your medication is safe for you" and "your medication will be restored if your symptoms worsen or if you request so" was used in the final version of the consent instead of "you might relapse" or "your psychosis might worsen." It is probable that the Internal Review Boards considered that talking to the patients about psychosis or schizophrenia might cause unnecessary anxiety, and therefore, would not be in the best interest of the patient. Although this approach might appear paternalistic by 1994 standards, protecting patients, psychiatric and medical, from "bad news" were accepted standards in 1979.[25]

While acknowledging past wrongdoing, Davidson's letter suggested that things had changed. Researchers were no longer lying to the mentally ill in this way. Yet a review of consent forms for post-1985 symptom-exacerbation studies, obtained through Freedom of

Information requests, reveals more of the same. None of the forms stated that the experiments were expected to make the patients worse.

In 1994, for instance, Dr. Adam Wolkin and his colleagues at the NYU School of Medicine reported in *Biological Psychiatry* that they had conducted an experiment in which they used PET technology to "evaluate metabolic effects in subjects with varying degrees of *amphetamine-induced psychotic exacerbation*" (italics added). However, they had told their patients a different story in the consent form. There, they had said that the experiment was designed "to measure any abnormalities in the activity of different parts of the brain using a procedure called Positron Emission Tomography and to relate these to the effects of certain medications on the symptoms you have." The "medication" that the NYU researchers were referring to was amphetamine, which, they told their patients, caused "most people" to "experience some feelings of increased energy and confidence." (They did note that there was a "risk" that amphetamine, in the manner of a side effect, might cause "some patients' symptoms" to "become more pronounced.")[26]

The consent form that Dr. Carol Tamminga and her colleagues at the University of Maryland used for their ketamine experiments was even more misleading. They expected that ketamine would both worsen their patients' delusions and blunt their emotions. However, in their consent form, they told patients that the experiment would "test a medication named ketamine for schizophrenia . . . this medication, if effective, may not alter [your] underlying disease, but merely offer symptomatic treatment." While they did acknowledge in the consent form that ketamine, when used as an anesthetic, had been known to cause "sensory distortion," they promised their patients that "at the dose levels which will be used in this study, no altered level of consciousness will occur."[27]

At a 1998 meeting held by the National Bioethics Advisory Commission, NIMH scientist Donald Rosenstein implicitly acknowledged that such obfuscation was routine. Researchers were not telling their patients that they were giving them chemicals *expected* to make them worse: "Everyone involved in these studies really needs to understand two things," he told the commission.

> One is that the purpose of the study is not to help. The purpose is to learn more about the underlying condition. The second—and this is also different than saying that this study may not be of benefit to you, which is typically how the language reads in a number of different consent forms—is that the symptoms are expected. They are not unintended side effects . . . I think a lot of people, including the investigators, can get confused about that.[28]

Even more telling was the reaction of ex-patients. When they learned about the experiments in 1998, they found them appalling in the extreme. They called them "evil," compared them to "Nazi experiments," and said they were reminiscent of abuses from "the psych wards of the gaslight era." "If a person is going through enormous suffering already, and then a doctor induces physical suffering on top of that, isn't that an abuse of power?" asked Michael Susko, who suffered a psychotic break at age twenty-five and works today with the homeless mentally ill in Baltimore.[29] Franklin Marquit, founder of the National Artists for Mental Health, surveyed a number of his fellow mental-health "consumers" on the topic and found that all objected vigorously to the studies, particularly to the notion that a transient worsening of symptoms posed little harm. "Have it done to yourself and see how the symptoms are," he said. "Someone who doesn't experience this traumatizing feeling, how would they know? With panic disorder, I feel like jumping off the edge of the earth at times, it is so bad."[30]

What bothered the ex-patients most of all, however, was the transparent hypocrisy of it all. "Their entire explanation is such horseshit," said Wesley Alcorn, president of the consumer council of the National Alliance for the Mentally Ill (NAMI) in 1998.

> Do you think people really say, "Gee, I'll sign up for more suffering?" Many of us suffer enough on our own. And these [researchers] are the same people who say we don't have enough insight and so there have to be involuntary commitment laws because we can't see that we are ill. Yet, now they say that we are well enough to agree to participate in these symptom-exacerbation studies, and that we are doing it of our own volition, and that society

> shouldn't deny us that right. The hypocrisy is mind-boggling. It shows that we are still dehumanized.[31]

Together, the paper trail and the reaction of ex-patients to the experiments point to one haunting conclusion. For fifty years, American scientists conducted experiments expected to worsen the symptoms of their mentally ill patients, and as they did so, time and time again they misled their patients, hiding their true purposes from them. This experimentation was done primarily on vulnerable people who did not know what was being done to them, which was precisely the type of science that the Nuremberg Code had sought to banish.

One American who can tell what it is like to be so misled and experimented on in this way is Shalmah Prince.

"I'll Never Be the Same"

Prince, who lives in Cincinnati, is a portrait artist. She graduated from Abilene Christian University in 1975 with a degree in fine arts, and then lived in New York City for a while, studying at the Art Students League and doing portraits for Bloomingdale's. In 1981, she suffered a manic episode and was diagnosed with manic-depressive (or bipolar) illness. Her doctors placed her on lithium, a medication that many patients find more tolerable than neuroleptics, but also one with a hidden cost. Patients who abruptly stop taking it are at high risk of relapse and may become sicker than they have ever been before. And if they do relapse, they might never quite fully recover, even after being placed back on lithium. Prince had done fairly well on the medication, but in early 1983, she started feeling edgy, and so she went to the emergency room at University Hospital in Cincinnati seeking help. She wanted to avoid another manic episode at all costs—her husband had left her during her first one.[32]

As her hospital records show, she arrived at the emergency room well groomed, alert, and thinking fairly clearly. The standard treatment, as Dr. David Garver and Dr. Jack Hirschowitz later admitted in court depositions, would have been to measure her lithium blood levels and then increase her medication to a therapeutic level, care that could have been provided on an outpatient basis. Instead, Prince was admitted to the hospital, and soon she found herself in a softly lit room, a staff doctor quietly asking if she'd like to be part of a research study. She would have to go without her lithium for a few days, she was told, and then she would be given a drug, apomorphine, expected to increase her human-growth hormone levels. The study, it seemed, was designed specifically to help a patient like her. The consent form she signed read: "I, Shalmah [Prince], agree to participate in a medical research study the purpose of which is to clearly diagnose my illness and determine whether treatment with lithium might provide long-term relief of my symptoms."

"I signed the form," Prince recalled. "I just wanted to be kept safe. I knew that I didn't have insurance and that I was extremely vulnerable. I needed help and a regular doctor was $150, so I was really stuck. You don't want to go manic. Besides, I was in a hospital, and I had this idea that when you went to a hospital and you had doctors seeing you that their purpose was to make you better. That's what they told me. They assured me they were there to treat me."

In fact, Prince was now a subject in an experiment on the "biology of schizophrenia subtypes" that would require her to *forgo* treatment. She would be kept off her lithium for at least a week, and then she would be injected with apomorphine, a dopamine-releasing agent that others had tested to see whether it would stir psychosis. It was a regimen that put her at high risk of suffering the manic attack she so feared. As Garver admitted in his deposition, the abrupt withdrawal of lithium medication could cause a bipolar patient "to have a delusion or otherwise act in irresponsible ways so as to harm themselves or someone else." The reason that the consent form didn't warn Prince of this risk, he said, was that "this risk would seem to be self-evident even to a person without medical training."

As could be expected, Prince's condition quickly deteriorated once her lithium was abruptly withdrawn. She grew louder and more boisterous, and she couldn't sleep at night. She joked with the nurses, saying, "I hope that the growth hormone you are giving me will make my breasts bigger"—a comment that showed she had little understanding of what the experiment was about. On January 17—her fourth day without lithium—her emotions careened totally out of control. She "got in the face" of another patient, and he started beating her. At some point, she set fire to some furniture, put a bag over her head, and threatened suicide. Hirschowitz judged her manic symptoms to have become "reasonably severe."

Even so, he still did not put her back on lithium.

Instead, on the morning of January 19, hospital doctors injected her with apomorphine. Her manic and delusional behavior quickly soared. "I was completely psychotic," she recalled. "I remember thinking that I could transfer myself to South America. I was totally afraid that I was losing my mind. And I was in a unit where everybody else had been injected and taken off medication. I was afraid for my life. We were begging for help, we were feeling so helpless." Prince's behavior deteriorated to such an extent that doctors slapped her into leather restraints. For three days she remained tied up like that, and while she was in that humiliating condition, bearing the dress of a madwoman, her family, friends, and boyfriend were allowed to visit, gaping in amazement at the sight of her.

> "After that, I was never the same person ever again," she says today.
>
> "I was so depressed and non-functioning, and confused and humiliated. Laying there in restraints, and having your family and friends and boyfriend see you—it was a total loss of dignity. You just lost it. By the time I left the hospital my perception of myself and who I was had completely changed. I had a sense of shame and embarrassment. It had changed my ability to relate socially. I had to start my friendships, my career plans, and even my idea of who I was kind of from scratch."

At the time, Prince had no idea what had happened to her. When she was released from the hospital, she was billed $15,000 for the "care" she'd received, and she focused on putting her ruined life back together. It wasn't until 1994, when she read an article in *U.S. News and World Report* about Cold War radiation experiments on unsuspecting Americans, that she suddenly wondered about her miserable experience years earlier. Had she too been used? Over the next few years, she painstakingly pieced together what had happened to her. She forced the hospital to produce her medical records and a copy of the research protocol she'd been in, and by suing the doctors, she got them to explain why they hadn't informed her of the risks. The record of deception was all there.

However, that perseverance led to a bitter end for Prince. The judge in her lawsuit, while finding the "facts" troubling, dismissed her case, ruling that, with due diligence, she could have learned at a much earlier date how she'd been used and thus should have properly filed her complaint within two years of the experiment, as required by the statute of limitations. The attorney for the doctors, Ken Faller, even suggested that Prince didn't have much to complain about in the first place: "She did receive treatment and the treatment benefited her to this day," he said. "She was a sick person when she went into the hospital and she came out seemingly in pretty good shape."[33]

Today, NIMH-funded symptom-exacerbation experiments appear to have ceased. The public spotlight that was shone on the experiments in 1998 caused NIMH to reconsider this line of research, and it subsequently halted a number of studies. As for the promised clinical advances, fifty years of experimentation brought none to fruition. The biology of schizophrenia is still not at all well understood, there is still no diagnostic test for schizophrenia, and the development of the new "atypicals" marketed in the 1990s cannot be traced to this research. There is not a single advance in care that can be attributed to a half century of "modeling psychosis" in the mentally ill.

PART FOUR

MAD MEDICINE TODAY

(1990s–Present)

11

NOT SO ATYPICAL

This is a field where fads and fancies flourish. Hardly a year passes without some new claim, for example, that the cause or cure of schizophrenia has been found. The early promises of each of these discoveries are uniformly unfulfilled. Successive waves of patients habitually appear to become more resistant to the newest "miracle" cure than was the group on which the first experiments were made.

—Joint Commission on Mental
Illness and Mental Health, 1961[1]

ONE OF THE enduring staples in mad medicine has been the rise and fall of cures. Rarely has psychiatry been totally without a remedy advertised as effective. Whether it be whipping the mentally ill, bleeding them, making them vomit, feeding them sheep thyroids, putting them in continuous baths, stunning them with shock therapies, or severing their frontal lobes—all such therapies "worked" at one time, and then, when a new therapy came along, they were suddenly seen in a new light, and their shortcomings revealed. In the 1990s, this repeating theme in mad medicine occurred once again. New "atypical" drugs for schizophrenia were brought to market amid much fanfare, hailed as "breakthrough" treatments, while the old standard neuroleptics were suddenly seen as flawed drugs, indeed.

However, there was something different about this latest chapter in mad medicine.

Prior to the introduction of chlorpromazine, belief in the efficacy of a treatment usually rose in a haphazard way. The inventor of a therapy would typically see it in a rosy light, and then others, eager for a new somatic remedy with which to treat asylum patients, would find it helpful to some degree. And all of the old therapies did undoubtedly "work." They all served to quiet or weaken patients in some way, and that was a behavioral change that was perceived as good. With chlorpromazine, the belief in efficacy was shaped for the first time by a well-organized company pursuing profits. Yet at that time, the pharmaceutical industry was still in its infancy, and the apparatus for weaving a story of a new wonder drug wasn't all that well developed. The transformation of chlorpromazine from a drug that induced a chemical lobotomy into a safe, antischizophrenic drug took a decade. But by the late 1980s, the pharmaceutical industry's storytelling apparatus had evolved into a well-oiled machine. The creation of a tale of a breakthrough medication could be carefully plotted. Such was the case with the atypicals, and behind the public facade of medical achievement is a story of science marred by greed, deaths, and the deliberate deception of the American public.

Recasting the Old

The atypicals were brought to market at a time when Americans had become ever more certain of the therapeutic efficacy of antipsychotic medications. The National Alliance for the Mentally Ill had grown up in the 1980s, and its message was a simple one: Schizophrenia is a biological disorder, one caused by abnormal chemistry in the brain, and medications help normalize that chemistry. That same basic paradigm was used to explain other mental disorders as well, and America—gobbling up antidepressants, antianxiety agents, and any other number of psychotropic medications—had in essence accepted it as a way to understand the mind. With this conception of mental illness at work, even patients' protests against neuroleptics dimmed. They apparently had broken brains and needed the drugs—however unpleasant they might be—to set their minds at least somewhat straight.

And so, as the atypicals arrived, two somewhat curious stories about the therapeutic merits of old neuroleptics were told—one for the ears of other doctors, and one for the ears of the public.

The selling of new drugs necessarily involves telling a story that contrasts the new with the old. The worse the old drugs are perceived to be, the better the new drugs will look, and so as the atypicals moved into the marketplace—which meant that drug firms were hiring well-known psychiatrists to serve as consultants and to run clinical trials—researchers started tallying up the shortcomings of standard neuroleptics. It was an exercise that even seemed to produce a momentary air of liberation within American psychiatry. For so long, investigators had held to the story that Thorazine, Haldol, and the others were effective antipsychotic medications, ultimately good for their patients, and now, at long last, they were being encouraged to see these drugs in an alchemy-free light.

The old drugs, researchers concluded, caused a recognizable pathology, which they dubbed neuroleptic-induced deficit syndrome (NIDS). As would be expected, NIDS was a drug-induced disorder that mimicked natural diseases—like Parkinson's or encephalitis lethargica—that damaged dopaminergic systems. Two-thirds of all drug-treated patients, researchers calculated, were plagued by "persistent Parkinson's." Nearly all patients—some physicians put the figure at 100 percent—suffered from extrapyramidal symptoms (EPS) of some type. (Extrapyramidal symptoms include all of the various motor side effects, such as Parkinson's, akathisia, and muscle stiffness.) As for tardive dyskinesia, investigators announced that it might be more of a risk than previously thought: It struck up to 8 percent of patients in their *first year* of exposure to a potent neuroleptic like haloperidol. The list of adverse effects attributed to neuroleptics, meanwhile, grew to head-spinning length. In addition to Parkinson's, akathisia, blunted emotions, TD, and neuroleptic malignant syndrome, patients had to worry about blindness, fatal blood clots, arrhythmia, heat stroke, swollen breasts, leaking breasts, impotence, obesity, sexual dysfunction, blood disorders, painful skin rashes, seizures, and, should they have any children, offspring with birth defects. "They have adverse side effect profiles that can affect every physiologic system," said George Arana, a psychiatrist at the Medical University of South Carolina, at a 1999 forum in Dallas. Nor was it just bodily functions so impaired. "Typical antipsychotic medications," Duke University's Richard Keefe told his peers, may "actually prevent adequate learning effects and worsen motor skills, memory function, and executive abilities, such as problem solving and performance assessment."[2]

Researchers also began to admit that neuroleptics didn't control delusions and hallucinations very well. Two-thirds of all medicated patients had persistent psychotic symptoms a year after their first psychotic break. Thirty percent of patients didn't respond to the drugs at all—a "non-response" rate that up until the 1980s had hardly ever been mentioned. Several studies suggested that even this 30-percent figure might be very low and that as many as two-thirds of all psychotic patients could be said to be "non-responders" to neuroleptics.[3] Perhaps the most revealing confession of all came from NIMH scientists: "Our clinical experience is that while the intensity of thought disorder may decrease with medication treatment, the profile of the thought disorder is not altered."[4] The drugs, it seemed, might not be "antipsychotic" medications after all.

As for the patients' quality of life, nearly everyone agreed that neuroleptics had produced a miserable record. More than 80 percent of schizophrenics were chronically unemployed. Their quality of life is "very poor," wrote New York's Peter Weiden. Said Arana: "Patients still lie in bed all day. They are suffering." Long-term outcomes with neuroleptics, commented Philip Harvey, from the Mt. Sinai School of Medicine in New York City, were no better than "when schizophrenia was treated with hydrotherapy." Said one physician at the Dallas conference: "We will do a great service to our [first-episode] patients by never exposing them to typical antipsychotic drugs." A 1999 patient survey completed the profile: Ninety percent on neuroleptics said they were depressed, 88 percent said they felt sedated, and 78 percent complained of poor concentration.[5]

All of this was undoubtedly quite true, and yet it had come at a telling time. New drugs were coming to market and such candor about the old ones served as a powerful foil for making the new ones look good. Psychiatrists who came to the Dallas conference, which was sponsored by Janssen, the manufacturer of the atypical drug risperidone, couldn't have missed the message: Those who tended to the severely mentally ill would do well to begin prescribing Janssen's new drug and other atypicals as quickly as possible. The financial forces that helped drive perceptions within psychiatry had changed, and that had led—*within* the medical community—to a rather stunning reassessment of the old.

But what to tell the public? Neuroleptics—billed as antipsychotic medications—had been the mainstay treatment for schizophrenia for forty years. Over and over again the public had been told that schizophrenia was a biological disease and that drugs helped alleviate that biological illness. The drugs were like "insulin for diabetes." What if psychiatry now publicly confessed that the dopamine theory hadn't panned out, that the drugs induced a disorder called NIDS, and that outcomes were no better than when the mad were plunked into bathtubs for hours on end? At least hydrotherapy hadn't caused tardive dyskinesia, Parkinson's, or a host of other side effects. What would the public make of that admission?

A subtler story emerged in public forums. The old drugs were beneficial, but problematic. The new drugs were a wonderful *advance* on the old. As for the tired dopamine theory, it too proved to have some life left in the public domain.

"Breakthrough" Treatments

From a business perspective, the introduction of a new antipsychotic medication was long overdue when the first atypical drug, clozapine, was brought to the U.S. market in 1990 by Sandoz. By the early 1980s, the market for neuroleptics had devolved into a relatively unprofitable phase. There were more than a dozen neuroleptics on the market, and the leading ones—chlorpromazine and haloperidol—had long lost their patent protection and thus were vulnerable to generic competition. Chlorpromazine was selling for less than $10 per month, and haloperidol for not a great deal more. Sales for all neuroleptics in the United States in the late 1980s totaled less than $400 million, which was much less than what one "breakthrough" medication could hope to generate in a year. The market was ripe for a novel antipsychotic, and it came in the form of a drug that, fifteen years earlier, had been discarded as too dangerous.

Clozapine, marketed by Sandoz as Clozaril, was first tested as an antipsychotic in the 1960s. It was different from other neuroleptics in that it blocked both dopamine and serotonin receptors. When tested, it was found that it didn't cause the usual high incidence of extrapyramidal symptoms. However, it did cause any number of other neurotoxic effects—seizures, dense sedation, marked drooling, rare sudden death, constipation, urinary incontinence, and weight gain. Respiratory arrest and heart attacks were risks as well. Sandoz introduced it into Europe in the 1970s, but then withdrew it after it was found to also cause agranulocytosis, a potentially fatal depletion of white blood cells, in up to 2 percent of patients.

The return of clozapine was made possible by the fact that, by the mid-1980s, it was no longer possible to ignore the many drawbacks of neuroleptics. Because of the risk of agranulocytosis, the FDA approved it only as a second-line therapy for patients who didn't respond to standard neuroleptics. Even so, it quickly proved to be a hit in the marketplace. It didn't appear to cause extrapyramidal symptoms, and at least some patients responded—in terms of the clarity of their thinking—in a robust fashion. Sandoz also initially sold clozapine bundled with weekly blood tests for agranulocytosis, with the test to be done by its affiliate, Caremark, and it put a whopping price of $9,000 a year on the package.

Other drugmakers now had a clear model to emulate. A drug that could block both serotonin and dopamine receptors could hopefully knock down psychosis without causing the usual extrapyramidal symptoms, and it might even improve cognition. Any drug that could do that without causing agranulocytosis could be marketed as a first-line therapy, and generate blockbuster financial returns. In the early 1990s, the medical literature began bubbling with reports of just such a drug, risperidone. Janssen obtained FDA approval in 1993 to sell it, and by the end of 1995, more than twenty reports had appeared in psychiatric journals touting its benefits. It was said to be equal or superior to haloperidol in reducing positive symptoms (psychosis), and superior to haloperidol in improving negative symptoms (lack of emotion). Researchers reported that it reduced hospital stays, improved patients' ability to function socially, and reduced hostility. Best of all—and this was the sound bite that graced journal advertisements—the incidence of extrapyramidal symptoms with risperidone was said to be "equal to placebo."[6]

The media presented risperidone in even more glowing terms. This new drug, the *Washington Post* reported, "represents a glimmer of hope for a disease that until recently had been considered hopeless." Risperidone, it said, did not "cause sedation, blurred vision, impaired memory or muscle stiffness, side effects commonly associated with an earlier generation of antipsychotic drugs." George Simpson, a physician at the Medical College of Pennsylvania, told the *Post:* "The data is very convincing. It is a new hope, and at this moment it appears, like clozapine, to be different from all existing drugs." The *New York Times,* quoting Richard Meibach, Janssen's clinical research director, reported that "no major side effects" had appeared in any of the 2,000-plus patients who had been in the clinical trials. The *Times* also provided its readers with a diagram of how risperidone worked. "Researchers," it said, think that drugs like risperidone "relieve schizophrenia symptoms by blocking excessive flows of serotonin or dopamine, or both."[7]

The dopamine theory, in a slightly amended version, was alive and well. Schizophrenics suffered from not just one neurochemical abnormality, but two, and the new atypicals helped normalize both. As for the older drugs, the *New York Times* reported, they "relieve typical symptoms like delusions and hearing voices in about 70 percent of patients. But they are less effective in treating other symptoms of schizophrenia, like withdrawal, lack of energy and motivation, and the inability to experience pleasure." All of the other papers cast the standard neuroleptics in that same light: They were *less effective* (or ineffective) in treating negative symptoms. They did successfully treat positive symptoms in about 70 percent of patients. None of the newspapers told of how the older drugs could impair cognitive function and worsen negative symptoms, nor was it mentioned that they caused a recognizable pathology, known as

NIDS, or that, as Philip Harvey had written, it might be that they "had no impact on the overall outcome of schizophrenia."[8] Instead, in this story told to the public, risperidone's arrival in the marketplace was successfully placed within the framework of the long-running story of the general *efficacy* of neuroleptics. The tale of helpful, antipsychotic drugs was maintained.

It was also a story that Janssen took to the bank. With praise flowing in the scientific literature and in the media, Janssen was able to charge $240 per month for risperidone, more than thirty times the price of chlorpromazine. In 1996, U.S. sales of risperidone topped $500 million, which was greater than revenues for all other neuroleptics combined. That same year, Janssen won the prestigious Prix Galien for its new drug, an award touted as the pharmaceutical industry's Nobel Prize.

Eli Lilly was the next to bring an atypical to market. However, since Janssen had made it first to the marketplace, Eli Lilly's challenge was to prove in clinical trials that its new drug, olanzapine (marketed as Zyprexa), was superior to both haloperidol and risperidone. Olanzapine was chemically more similar to clozapine than Janssen's drug (risperidone blocked D_2 receptors in a more potent manner than did clozapine or olanzapine), and as olanzapine came to market in 1996, reports in the medical journals told just the story that Eli Lilly wanted. Olanzapine, the articles said, worked in a more "comprehensive" manner than either risperidone or haloperidol. It was a well-tolerated agent that led to global improvement—it reduced positive symptoms, caused fewer motor side effects than either risperidone or haloperidol, and improved negative symptoms and cognitive function. It reduced hospital stays, prevented relapse, and was useful for treatment-resistant schizophrenia.[9]

Apparently, yet another step up the medical ladder had been taken. Olanzapine, the *Wall Street Journal* announced, has "substantial advantages" over other current therapies. "Zyprexa is a wonderful drug for psychotic patients," said John Zajecka, at Rush Medical College in Chicago. Harvard Medical School's William Glazer told the *Wall Street Journal:* "The real world is finding that Zyprexa has fewer extrapyramidal side effects than Risperdal." Stanford University psychiatrist Alan Schatzberg, meanwhile, confessed to the *New*

York Times: "It's a potential breakthrough of tremendous magnitude." On and on it went, the glowing remarks piling up. Laurie Flynn, executive director of the National Alliance for the Mentally Ill, even put an exclamation point on it all: "These new drugs truly are a breakthrough. They mean we should finally be able to keep people out of the hospital, and it means that the long-term disability of schizophrenia can come to an end."[10]

Since its drug was seemingly better than Janssen's, Eli Lilly was able to put a higher price tag on it. Patients would have to pay nearly $10 per day for this new miracle drug. In 1998, olanzapine sales in the United States alone topped $1 billion. Total U.S. sales of antipsychotic drugs hit $2.3 billion that year—roughly six times what they had been prior to risperidone's arrival on pharmacy shelves. By that time, AstraZeneca had brought a third atypical to market, quetiapine (marketed as Seroquel), and there was no longer any possible doubt about the superiority of these new drugs. They were, *Parade* magazine told its readers, "far safer and more effective in treating negative symptoms, such as difficulty in reasoning and speaking in an organized way." The *Chicago Tribune* echoed the sentiment: The newer drugs "are safer and more effective than older ones. They help people go to work." Or as the *Los Angeles Times* put it: "It used to be that schizophrenics were given no hope of improving. But now, thanks to new drugs and commitment, they're moving back into society like never before."[11]

American science had surely produced a remarkable medical advance. New wonder drugs for madness had arrived.

The Business of Clinical Research

This belief—that the atypicals were superior in safety and efficacy—had a solid scientific pedigree. It was based upon the results of the clinical trials that the pharmaceutical companies had conducted to gain FDA approval for their drugs, which had been published in the best peer-reviewed medical journals. The *American Journal of Psychiatry*, *Journal of Clinical Psychopharmacology*, *Neuropsychopharmacology*—the literature was filled with articles praising the drugs. They were authored by some of the leading lights in American psychiatry, and inevitably the articles included an impressive array of statistics and charts, detailed explanations of methodology, and sober-minded conclusions. What the public couldn't have known is that this whole arena of science—the clinical testing of drugs—had undergone a profound change in the 1990s, one that lent itself to the creation of fairy tales, and that the FDA, in its review of the same trial data, didn't buy the companies' claims of superiority at all.

The refashioning of the clinical testing of commercial drugs can be traced back to the mid-1980s. Up until that point, pharmaceutical firms primarily hired academic physicians to test their drugs. More than 70 percent of all drug trials were conducted in academic settings, and the relationship between the drug companies and the academic doctors was one in which the doctors, in many ways, had the upper hand. The academic physicians often viewed the drug companies with more than a little disdain—grants from the National Institutes of Health were the coveted coin in the academic realm—and the drug companies basically had to come to the physicians as humble supplicants. The academic doctors were known as Thought Leaders, and the fact that they had the upper hand in the relationship ensured that experimental drugs went through at least a measure of independent testing. The academic doctors regularly modified the protocols, even though this often greatly irritated the drug companies.

However, starting in the late 1980s, a for-profit clinical trials industry arose to *serve* the pharmaceutical companies. It emerged in bits and pieces. First, community physicians who were feeling financially squeezed by health maintenance organizations turned to clinical trials as a way to supplement their incomes. Some conducted trials as an adjunct to their regular practices, while others opened full-time "dedicated" research centers. Then a group of urologists, from nineteen states, banded together to form Affiliated Research Centers. A pharmaceutical company developing a urology drug could come to Affiliated Research Centers and immediately have community physicians across the country lined up to test it. Doctors in other specialties soon established similar investigator networks. Next came pure business ventures, eager to consolidate services for the pharmaceutical firms. Entrepreneurs raised venture capital with the goal of building nationwide chains of research sites. By 1997, venture capital groups had poured $100 million into such businesses, and two of these venture-funded companies had turned public. It all led Peter Vlasses, director of clinical research for a consortium of university hospitals, to lament: "Everybody under the sun is now a clinical researcher. What used to take place only in academic centers is now everywhere."[12]

As this mix of for-profit research sites sprung up, spending by pharmaceutical companies for their services soared, from under $1 billion in 1990 to $3.5 billion in 2000. The role of these for-profit businesses in the research process was very straightforward: Their job was to recruit patients quickly into trials and keep them there until they completed the study protocols. Said one Texas investigator in 1995: "I don't begrudge [the pharmaceutical companies] viewing me as a vendor. I am providing a technical service, and in that sense, I view it as a business. If I were not turning a profit, I wouldn't do it. And I don't think many investigators would." There certainly was money to be made. In 1997, community physicians experienced at conducting clinical trials reported earning, on average, $331,500 from their research activities. "Dedicated" research centers reported revenues of $1.35 million. A newsletter for neurologists, *Neuropractice*, summed up the opportunity in commercial drug trials: "A growing number of neurologists are discovering a gold mine in their clinical practices: their patient population." A few investigators chalked up even bigger scores. In 1996, pharmacist Jeff Green took his company, Collaborative Clinical Research, public, raising $42 million for his expansion plans. Two Rhode Island psychiatrists, Walter Brown and Michael Rothman, reaped the biggest financial success of all. In 1997, they sold their seven-year-old company, Clinical Studies, which consisted of a chain of research centers along the East Coast, for stock valued at $96 million.[13]

The commercial testing of experimental drugs had moved out of an academic setting and into a for-profit setting. Struggling to cope with this loss of business, academic centers also began changing their ways. A number of schools opened administrative offices devoted to securing contracts for commercial drug trials. The central "offices of clinical trials" promised the pharmaceutical firms that they would help their physicians start trials quickly and successfully fill them with patients. They too adopted a *service* attitude toward the drug firms—that's what it now took to compete in the clinical-trials business. And with the old disdain toward pharmaceutical money melting away in academia, individual faculty became more eager to work for the drug firms as well. In a 2000 editorial titled "Is Academic Medicine for Sale?" *New England Journal of Medicine* editor Marcia Angell catalogued the many ways that drug money flowed to academic doctors:

> The ties between clinical researchers and industry include not only grant support, but also a host of other financial arrangements. Researchers also serve as consultants to companies whose products they are studying, join advisory boards and speakers' bureaus, enter into patent and royalty arrangements, agree to be the listed authors of articles ghostwritten by interested companies, promote drugs and devices and company-sponsored symposiums, and allow themselves to be plied with expensive gifts and trips to luxurious settings. Many also have equity interest in the companies.[14]

In this new service environment, the drug companies enjoyed the best of all possible worlds. They could utilize for-profit research sites to recruit the bulk of their patients into their large clinical trials. At the same time, they could hire academic doctors to lend intellectual prestige and an aura of independence to the trial results. Together, these services produced the perfect package. The pharmaceutical companies could get their trials done quickly, the public would see the names of the academic physicians on the published articles, and all the while, they would control every aspect of the drug-testing process. They could, for instance, design their protocols without having to worry that academic doctors would insist on changing them, and that meant that it would now be easier for them to set up trials biased toward their own drugs. "A pharmaceutical company," acknowledged *Journal of Clinical Psychiatry* editor Alan Gelenberg in 1999, "goes to great pains to construct studies that are likely to turn out in its favor."[15] The drug companies also controlled analysis of the data, and that control, the *New England Journal of Medicine* wrote, "allows companies to provide the spin on the data that favors them."[16]

In short, a dark truth became visible in American medicine in the 1990s. Bias by design and the spinning of results—hallmarks of fraudulent science—had moved front and center into the testing of commercial drugs. While this corruption of the drug-testing process was not unique to psychiatry, it was no accident that the *New England Journal of Medicine,* as it sought to illustrate the problem, found the best evidence of it in this specialty. When the journal tried to identify an academic psychiatrist who could write an honest review of antidepressant drugs, it found "very few who did not have financial ties to drug companies." One author of an article on antidepressant drugs had taken money from drug companies on so many occasions, Angell told an ethics conference in 2000, that to disclose all of them "would have taken up more space than the article." She concluded: "You are seeing played out in psychiatry the extremes of what is happening elsewhere in medicine."[17]

And all of these extremes were at work as the atypicals came to market.

Eye on the Castle

One of the first academic physicians to tout the benefits of risperidone, in a 1992 article published in the *Psychopharmacology Bulletin,* was a psychiatrist at the Medical College of Georgia, Richard Borison. His 1992 report came to be frequently cited in the scientific literature, and over the next five years, he regularly published additional articles related to the merits of the atypicals. In 1994, he traveled to Australia to speak about risperidone, and he also was one of the experts quoted by the newspapers. Risperidone, he told the *New York Times* in 1992, was producing results that were "absolutely on the money."[18]

It was a quote that revealed more about Borison than it did about risperidone.

Although Borison was popular with the drug companies, he had a shady track record. In 1984, Smith Kline had given him a grant to conduct a test comparing Thorazine to a generic knockoff—in such studies, the drug company hopes to prove that the generic is not really equivalent—and the next year, at the American Psychiatric Association's annual convention, he reported the results that Smith Kline wanted to hear. Schizophrenics who had been switched from Thorazine to generic chlorpromazine had become agitated and hostile, he told his peers. His findings, which suggested that hospitals and caregivers would be wise to avoid generic chlorpromazine and buy Smith Kline's Thorazine instead, were widely circulated. However, the FDA chose to investigate his study, which Borison had conducted at Veterans Affairs Medical Center in Augusta in May 1984, and determined that the hospital hadn't even stocked Thorazine at that time. The patients could not have been switched from Thorazine to generic chlorpromazine at all—they had been on the generics all along. Although he tried to explain this damning finding away, the conclusion was obvious: Borison had simply fabricated the results.[19]

The FDA publicly rebuked Borison, but since he hadn't submitted his data to the agency, it lacked authority to formally discipline him. In the wake of the scandal, Borison's research activities lagged for a year or two, and then all was forgotten. He became a full professor at the Medical College of Georgia in 1988, was made chief of psychiatry at the VA hospital, and soon he and his research partner, Bruce Diamond, a pharmacologist on the faculty, had drug companies giving them one lucrative contract after another. Eli Lilly, Janssen, Zeneca, Sandoz, Glaxo, Abbott, Pfizer, Hoechst Marion Roussel—they all came knocking. The two researchers secured 160 contracts from drug firms over the course of a decade, worth more than $10 million. They received $4 million for schizophrenia drug trials alone. "We knew how to collect the information the way they wanted us to," one of Borison's employees told VA hospital officials in 1996. "And we were high enrollers [of patients into trials], so they loved us."[20]

As faculty, Borison and Diamond were supposed to get approval from the medical school to do drug studies. Payments for commercial trials were supposed to be sent directly to the school. But starting in 1989, Borison and Diamond cut the college out of the loop and told the drug firms to send their money directly to them. They opened an office across the street from the medical school and turned it into a full-time research mill, which they called Clinical Therapeutics. In order to keep the school in the dark about their research activities, they used a commercial service to do ethical reviews of their studies. The one thing they let the medical school continue to do was pay some of their expenses—they even placed Clinical Therapeutics' staff on the school's payroll.

To run their trials, Borison and Diamond hired attractive young women as study coordinators. When women came to apply for a coordinator's position, Diamond would wait to "see what they looked like in the waiting room," employee Angela Touhey told VA officials. "If they were overweight, if they were older, he would refuse to see them. He would ask a coordinator to talk to them and they would be sent home." There was a financial logic to their hiring preferences. The majority of patients recruited into trials are men, and that is particularly true of schizophrenia trials, which are among the best-paying studies in the business. Borison and Diamond stood to receive $10,000 to $25,000 for every schizophrenic the young women could coax into a drug trial.

With such money waiting to be made, Borison and Diamond gave the coordinators patient-recruitment bonuses that ran into the thousands of dollars. One coordinator was given a new Honda Accord as a bonus. Each time a new contract from a drug firm came in, the coordinators would hit the phones. They would call mentally ill people living in the community and promise them $150 if they would participate in the study. Patients already on locked wards at the hospital would be given cigarettes for participating. Some patients were churned through study after study, as well. "When there is a possibility you're going to get a car, you're going to do whatever you can," Touhey said.

Even though the coordinators lacked medical training, they regularly decided whether patients qualified for the trials. At times, they fudged information about the patients so that they met eligibility criteria. They also drew blood samples and adjusted the patients' drug dosages. Borison, employees said, rarely bothered to show up at the office. The coordinators would fill in the paper documents and then pass them on to Diamond, who would forge Borison's signature. At one weekly staff meeting, Touhey told the VA investigators, Diamond made it clear that he wasn't interested in hearing about the patients. "Bruce said to me, 'We don't care about how the patients are doing. We just want to know how many people you have enrolled in the past week or couple of weeks.'" Indeed, Borison and Diamond "had no idea who the patients were," Touhey said.

The money rolled in. Borison and Diamond stashed more than $5 million in cash and securities in various U.S. banks and Barclay's Bank in London. Each tooled around town in a new Mercedes Benz, and Diamond liked to show off his $11,000 gold Baume Mercier wristwatch. Borison's material dreams were even grander. He had an architect draw up plans for an 11,000-square-foot castle, complete with moat and medieval pennants. In anticipation of his new home, he made himself a regular at Sotheby's auction house, both in New York and London, purchasing such items as fifteenth-century tournament armor ($6,600), bronze doors ($16,000), a stone lion fountain ($32,000), two seven-foot stone entry lions on pedestals ($10,500), a marble statue of Cupid ($6,250), a crystal chandelier ($5,000), a coat of arms ($1,650), and more than 100 other decorative pieces—expensive paintings, marble vases, and antique furniture—that would make a castle fit for a king.

This went on for years. In early 1994, a study coordinator, Terri Davis, threatened to blow the whistle after a patient, who had been improperly admitted to an olanzapine trial, attempted suicide, but Borison and Diamond bribed her to keep quiet. A steady stream of monitors sent by the drug companies to audit their research records came and went. Borison would come in on the days the monitors were there and "set up a mock office," and the monitors would leave none the wiser. Risperidone was approved by the FDA in 1993, and the staff at Clinical Therapeutics even felt a measure of pride at their contribution—Borison had been a lead investigator in both of the pivotal U.S. studies Janssen had conducted. Finally, in 1996, Angela Touhey left and went to work for David Hess, chief of neurology at the Augusta VA hospital, and that triggered the collapse of Clinical Therapeutics. She told Hess about what had been going on, he investigated, and soon the police had been called. "This whole thing was very dirty," Hess told the medical school and hospital. "It was basically a numbers game. These patients are purely used for the greed of the researchers. That was very apparent to me what was going on."

Both Borison and Diamond went to prison, but not for research fraud. Their principal crime was that they had stolen from the college. Diamond was sentenced to five years in prison, fined $125,000, and ordered to pay $1.1 million to the college. Borison got fifteen years, was fined $125,000, and was ordered to pay $4.26 million to the college. As for his public comments about the merits of atypicals, Borison's last published article on the drugs—his eleventh overall—appeared in early 1997, about the same time that he was indicted. It was titled, "Recent Advances in the Pharmacotherapy of Schizophrenia," and it took him a full sixteen pages to detail all that he knew about how they had helped his patients get well.[21]

Swept Under the Rug

While the misdeeds of Borison and Diamond do not reveal anything about the merits of the atypicals, they do reveal much about the amount of money that was flowing to investigators who conducted the trials and how an academic physician who spoke well of a drug could expect a steady flow of research contracts, and polish up his CV at the same time. Their scandal provides insight into the *storytelling* forces at work as the new atypicals came to market. Those same forces can also be seen in a second behind-the-scenes aspect of the atypical trials, and that is how investigators reported on patient deaths. One in every 145 patients who entered the trials—for risperidone, olanzapine, quetiapine, and a fourth atypical called sertindole—died, and yet those deaths were never mentioned in the scientific literature.[22] Nor did anyone dare confess that the high death rate was due, in large part, to study design.

Pharmaceutical companies developing new drugs always want to get their trials done as quickly as possible. The adage in the industry is that every day delayed in the trial process is a million-dollar loss in potential sales. To get their atypicals approved, Janssen, Eli Lilly, and other companies needed to prove that the drugs reduced psychotic symptoms. Thus, they needed patients who were actively psychotic. To develop this patient pool (and do so quickly), they relied on protocols that required patients to be *abruptly* withdrawn from their existing medications. This abrupt withdrawal (also known as a "washout") could be expected to trigger a return of their hallucinations and delusions. Once the patients were newly sick, they could then be randomized into the trial and treated either with placebo, a standard drug like haloperidol, or the experimental drug. "If you don't take people who have reestablished active disease, then you don't know what you are looking at" when you test the drug, explained Robert Temple, director of the FDA's Office of Drug Evaluation. "That is why you have to have a washout."[23]

However, abrupt withdrawal (as opposed to gradual withdrawal) is also known to put patients at risk of severe clinical deterioration. It is contrary to good clinical practice and it increases the risk of suicide, which is precisely how many people died in the trials. At least thirty-six people in the studies of the four drugs killed themselves. Hanging, drowning, gunshots to the head, and death by jumping were some of the ways they chose to go. The overall suicide rate for patients in the trials, on a time-adjusted basis, was two to five times the norm for schizophrenics.*

One of the thirty-six people who died in this manner was forty-one-year-old Susan Endersbe, from Minneapolis. Her struggles with schizophrenia were of a familiar kind. She'd first begun to grapple with emotional difficulties as a teenager, and then she'd become more seriously ill while a student at the University of Minnesota. For the next twenty years, she went through many ups and downs. At times, she was able to live in her own apartment, with support from social services, but then her symptoms would worsen,

*In the medical literature, researchers report annual suicide rates for schizophrenics at two to five deaths per 1,000 people. In the atypical trials, the annual suicide rate for patients (on a time-adjusted basis) was close to ten per 1,000 people, or two to five times the norm. The number of patients in the research trials who committed suicide was also undoubtedly higher than thirty-six; dropout rates in the trials were quite high and many of these patients simply dropped off the researchers' radar screens.

and she would check herself into a hospital. The one constant was that she showed a will to live. "She was extremely intelligent and very high functioning for having such a disability, and recognized the illness for what it was," said her brother, Ed Endersbe. "She wanted very much to live and be a survivor."[24]

On May 7, 1994, she checked herself into Fairview Riverside Hospital in Minneapolis. It was a particularly difficult time for her—her mother had been diagnosed as terminally ill with cancer, and now Susan was feeling suicidal. Hospital doctors put her on an antidepressant, and gradually her mood lightened. On May 26, she told nurses that she was feeling much better and would be ready to leave soon. But the very next day, she was referred to psychiatrist Faruk Abuzzahab, and he had a different proposition for her. Would she like to be in a trial for a new drug, sertindole?[25]

Abuzzahab was a prominent psychiatrist in Minnesota. He'd served a term as president of the Minnesota Psychiatry Society and had chaired its ethics committee. He was also well known in the world of commercial drug research. He'd done a number of studies for pharmaceutical firms and had been a named author on published results. In the spring of 1994, he had a contract with Abbott Laboratories to test sertindole. However, the protocol specifically excluded patients who were suicidal. Nursing notes, according to her brother Ed, also showed that Susan Endersbe had reservations about entering a drug experiment. But no matter. On May 27, the day that Abuzzahab met Endersbe, he enrolled her in the study.

As the protocol stipulated, Abuzzahab immediately withdrew her medications. He also took her off the antidepressant venlaxafine, which had seemed to help her, and very shortly she began to deteriorate. Her emotional despair returned, and to make matters worse, she suffered a flare-up of extrapyramidal symptoms, a common occurrence when antipsychotic drugs are abruptly withdrawn. By June 3, nurses were writing that her suicidal feelings had returned. Devils were now struggling for her mind, her brother said. Even so, Abuzzahab kept her in the study, and on June 8, he randomized her into one of the study arms. She was, Abuzzahab wrote in research documents, experiencing "0" adverse events.

Nursing notes, however, told a different story:

> June 8: Passive thoughts of suicide with hopeless/helplessness in coping with changes from study. Patient feels hopeless, has suicidal thoughts of leaving the unit and jumping off the bridge on Franklin Ave.
> June 9: Patient states she feels suicidal and has been actively thinking about suicide, stating that she's different from others because when she attempts, she will succeed. Refuses to divulge method she has planned, however states she is unable to use the method while hospitalized. States she can agree to not harm self while in hospital.

On June 10, Susan Endersbe asked Abuzzahab for a day pass. The protocol prohibited patients from leaving the hospital during the first four weeks of the study, but Abuzzahab shrugged off this rule and granted her a pass for the next day. He didn't even require that anyone go along.

The next morning, Susan Endersbe prepared to go out. She took the time to dress neatly and to do her hair in a French braid. It was as though she were preparing for an event and wanted to look nice. She went to her apartment, where she watered her plants and gathered up a few keepsakes. As she left, she slipped the key back under the door. She would not be needing it any more—the thoughtful thing to do would be to leave it for the landlord. She then walked directly to the Franklin Avenue Bridge, which spanned the Mississippi River. Just as she had said she would, she clambered over the railing and leaped to her death.

"For nearly 20 years, my sister was managing to win the battle for her survival, and when she went on a drug study there were supposed to be safeguards in place to protect her," said her brother. "Not only were they not in place, they neglected to have the usual safeguards that she would have had if she stayed on as an inpatient in the hospital. And to wash people out from their medication, to take away any kind of treatment, that to me is inhumane. If they did that to someone with a physical illness, I would think it would be criminal."

All told, seven people killed themselves in the sertindole trials. At least ten patients did so in the risperidone trials, fifteen in the olanzapine studies, and four in the quetiapine experiments. They were casualties of a drug-testing process that required that "active disease" be reestablished in patients, but when it came time to report the trial results in the scientific journals, this loss of life was conveniently forgotten.*

Heart of Darkness

Borison's misdeeds, unreported patient suicides, Abuzzahab's callous neglect of Endersbe—all of these are dark splotches on the research landscape. They also lead, in stepping-stone fashion, to a much larger story, and that is how the trial process, in the case of the atypicals, was employed not to inform, but to mislead. This story is revealed in FDA documents obtained through Freedom of Information requests.

The scientific background to the clinical trials of the atypical drugs is, in some ways, a confusing one. On the surface, the trials appeared to straightforwardly compare the atypicals to placebo and to haloperidol. But surface appearances can be deceiving. In the first place, there was no true placebo group in the trials. The same "abrupt withdrawal" design that put patients at great risk also produced a placebo group that could be expected to fare poorly. The placebo group consisted of patients going through an event—abrupt withdrawal—that could be expected to make them worse, and then they were left untreated for that withdrawal-induced illness. While that trial design provided companies with a

*The Minnesota Board of Medical Practice suspended Abuzzahab's license in 1997 for his "reckless" treatment of Endersbe and other psychiatric patients. However, Morris Goldman, associate professor of psychiatry at the University of Chicago School of Medicine, who investigated the case for the Minnesota licensing board, believes that Abuzzahab's case raises broader questions about the integrity of commercial drugs studies. "What is the value of the data obtained in these trials?" he said, in an interview. "Abuzzahab would have the patient's diagnosis called one thing in the regular medical chart, and then the person would be put on a drug study and the person's diagnosis would be called something else to fit the criteria of the drug study. Then (during the study) he would say that the patients were improving, when the whole staff was saying that they were falling apart. The problem, as was seen with Abuzzahab, is that you don't know if the data was fudged."

convenient placebo foil for making their drugs look good, it made for poor science. Harvard Medical School's Ross Baldessarini put it this way: "It could exaggerate drug-placebo differences, and you could get a stronger impression of the benefit of the drug. It may not be a completely fair comparison."[26] In the second place, as the FDA reviewers repeatedly pointed out, Janssen and Eli Lilly used biased trial designs to favor their experimental drugs over the standard neuroleptics.*

Janssen's risperidone was the first of the three atypicals (excluding clozapine) to undergo FDA review. The company conducted three "well-controlled" trials to support its New Drug Application.[27]

In the first, involving 160 patients at eight U.S. centers, risperidone was compared to placebo. Nearly 50 percent of the risperidone patients didn't complete the six-week trial. Risperidone was superior to placebo in reducing positive symptoms, but neither risperidone nor haloperidol was superior to placebo on the "Clinical Global Impression Scale," which measures overall improvement.

In the second, which involved 523 patients at twenty-six sites in the United States and Canada, four doses of risperidone were compared to a 20-milligram dose of haloperidol and to placebo. Forty-five percent of the risperidone-treated patients didn't complete the eight-week trial. Janssen maintained that this study showed that risperidone, at an optimal dose of 6 milligrams, was superior to haloperidol for treating positive and negative symptoms, which were the conclusions published in the medical journals. However, FDA reviewers noted that Janssen had used a single, high dose of haloperidol for comparison, a dose that "may have exceeded the therapeutic window" for some patients, and thus the study was

*None of the drug companies needed to prove their drugs were superior to standard neuroleptics in order to gain approval. They simply had to show that their experimental drugs reduced psychotic symptoms over a short period more effectively than "placebo." This was the "efficacy" requirement. To pass the safety hurdle, the drug companies primarily had to show that their atypicals didn't carry a high risk of death from side effects, such as cardiac problems. The drugs could cause an array of nonfatal side effects (extrapyramidal symptoms, and so on) and still gain approval. Such risks would simply have to be mentioned in warnings on the label.

"incapable by virtue of its design of supporting any externally valid conclusion about the relative performance of haloperidol and Risperdal."

This second risperidone trial conducted by Janssen clearly illustrates how trial design can be used to produce results a company wants. Haloperidol was a drug that had been in widespread use for more than twenty years, and it was well known—as the FDA reviewers pointed out—that high doses were problematic. For instance, Theodore van Putten at UCLA had reported in 1987 that a 20-milligram dose of haloperidol was "psychotoxic" to many patients and that even a 10-milligram dose triggered painful akathisia in 76 percent of patients. Similarly, in 1991, Duke University researchers determined that doses of haloperidol above 10 milligrams daily regularly led "to significant increases in distressing extrapyramidal side effects."[28] By using a 20-milligram dose, then, Janssen could expect that there would be a high incidence of extrapyramidal side effects in the haloperidol group and thus help create a story of how risperidone, by comparison, was a much safer drug.

In its third study, which involved 1,557 patients in fifteen foreign countries, Janssen compared five doses of risperidone to a 10-milligram dose of haloperidol. Janssen claimed that this study showed that its drug was "more effective than haloperidol in reducing symptoms of psychosis," but Paul Leber, director of the FDA's Division of Neuropharmacological Drugs, once again rejected this argument. The study was "incapable" of making any meaningful comparison. The design flaw in this study, Leber noted, was that Janssen had compared multiple doses of its experimental drug to one dose of haloperidol. In order to honestly compare two drugs, an equal number of "equieffective" doses must be tested, as otherwise the study unfairly favors the drug that is given in multiple doses. Such trial design, Leber wrote on December 21, 1993, is "a critical preliminary step to any valid comparison of their properties."[29]

In sum, the FDA concluded that Janssen had shown evidence that risperidone was effective in reducing positive symptoms compared to placebo over the short term but had not proven that its new drug was superior to haloperidol (which wasn't required for approval). As for risperidone's safety profile, a review of the FDA data shows it was much more problematic than the public had been led to believe. Researchers had proclaimed that the incidence of extrapyramidal symptoms was the "same as placebo." The *New York Times,* quoting a Janssen official, had reported that "no major side effects" had occurred in 2,000-plus patients. Those were results that spoke of a very safe drug. In fact, eighty-four risperidone patients—or about one in every thirty-five—had experienced a "serious adverse event" of some type, which the FDA defined as a life-threatening event, or one that required hospitalization. (Suicides and suicide attempts accounted for more than half of these serious events.) Moreover, in general, the incidence of adverse events in risperidone patients and haloperidol patients was roughly the same. Nine percent of risperidone patients had to drop out because of adverse events, compared to ten percent of haloperidol patients. Seventy-five percent of risperidone patients experienced at least one adverse event, compared to 79 percent of haloperidol patients. Even on a moderate dose of risperidone, 17 percent of risperidone patients suffered extrapyramidal symptoms, and at a high dose, one-third of risperidone patients did—which was about the same incidence of EPS in patients treated with 20 milligrams of haloperidol.* Wrote FDA scientist Thomas Laughren: "It remains to be seen how risperidone compares with other antipsychotics with regard to EPS, as haloperidol is at the high end of the spectrum."

In its final letter of approval to Janssen, the FDA made explicit its conclusions about the relative merits of risperidone and haloperidol. Robert Temple, director of the FDA's Office of Drug Evaluation, told Janssen:

---

*The trials clearly showed that EPS was a common risk with risperidone. The reason that Janssen could claim that extrapyramidal symptoms with moderate doses of risperidone were no worse than placebo was precisely because there was no real placebo group in the trials. In the Janssen trials, about one in six "placebo" patients experienced extrapyramidal symptoms. The symptoms are a *drug-withdrawal* effect, and not due to the disorder. The incidence of EPS in patients who received a fairly low dose of risperidone, 6 mg., was approximately the same. Thus, Janssen could claim that its drug caused EPS no more often than placebo did, which, to a naive public, suggested that it was risk free in this regard. While it was ludicrous science, it proved to be effective marketing.

> We would consider any advertisement or promotion labeling for RISPERDAL false, misleading, or lacking fair balance under section 502 (a) and 502 (n) of the ACT if there is presentation of data that conveys the impression that risperidone is superior to haloperidol or any other marketed antipsychotic drug product with regard to safety or effectiveness.[30]

However, while the FDA had the authority to stop Janssen from making false claims in its ads, it had no control over what academic physicians, who had been paid by Janssen to conduct the trials, reported in their medical journals or told the press. They had touted risperidone as superior to haloperidol prior to the FDA's review of the data, and they continued to do so afterward. In 1997, a group of elite academic psychiatrists revisited the trial data one last time, and in the *Journal of Clinical Psychiatry,* they once more told the story of its superiority. They wrote: "Our findings suggest that risperidone has important advantages compared with haloperidol. When administered in an effective dose range, risperidone produced greater improvements on all five dimensions of schizophrenia."[31]

In modern American psychiatry, the scientific journals had become a place to make claims that the FDA had explicitly banned from advertisements as false.

The FDA, however, had simply critiqued Janssen's trials as biased—it didn't conduct its own studies on the relative merits of risperidone and haloperidol. But once risperidone was on the market, physicians who had not received any money from Janssen could get their hands on the drug and conduct their own studies, and their results revealed, in dramatic fashion, just how egregiously the public had been misled, particularly in regard to the company's claims that extrapyramidal symptoms were the "same as placebo."

First, physicians at McMaster University in Hamilton, Ontario, found that in a study of 350 patients never before treated with neuroleptics, a low dose of risperidone caused Parkinsonism in 59 percent of the patients, compared to 52 percent of patients treated with haloperidol. The incidence of akathisia was also higher in the risperidone patients, leading the researchers to conclude that "risperidone may not be a useful alternative to typical antipsychotic drugs."[32]

Second, NIMH researchers determined that when risperidone and haloperidol were compared at equivalent therapeutic levels, risperidone induced extrapyramidal symptoms in 42 percent of the patients, compared to 29 percent in the haloperidol group.[33]

Third, University of Pittsburgh researchers determined that risperidone, when administered to neuroleptically naive patients, caused a disruption in eye movement still present four weeks after treatment was initiated, evidence of a neurological side effect lingering for a much *longer* time than it did in patients treated with haloperidol.[34]

Those studies were just the beginning of reports that, unbeknownst to the public, stripped much of the "breakthrough" luster from risperidone. In 1995, physicians at the University of Pittsburgh Medical Center complained that while the hospital's spending on antipsychotic medications had soared after risperidone was introduced, it couldn't find evidence that the drug produced better outcomes. Psychiatrists at the University of California at San Francisco, meanwhile, determined that only 29 percent of patients initially placed on risperidone were still on the drug two years later, with 55 percent quitting the drug because it didn't work. "Our findings suggest that in routine clinical practice, use of risperidone is plagued by many of the same problems that are well known with older antipsychotic medications," they wrote. Yet another researcher, Jeffrey Mattes, director of the Psychopharmacology Research Association, concluded in 1997 that "it is possible, based on the available studies, that risperidone is not as effective as standard neuroleptics for typical positive schizophrenia symptoms." Letters also poured in to medical journals linking risperidone to neuroleptic malignant syndrome, tardive dyskinesia, tardive dystonia, liver toxicity, mania, and an unusual disorder of the mouth called "rabbit syndrome."[35] A final blow was delivered in the prestigious medical journal *Lancet.* Janssen's clinical investigators had published results from the same trial multiple times, and critics held up this behavior as illustrative of the "salami science"—characterized by "redundant publication, slippery authorship, and opaque reporting of trial data"—that was poisoning the medical literature. Risperidone, one *Lancet* writer snapped, was "a marketing success, if nothing else."[36]

But the public heard little of this. The FDA's criticisms took place behind closed doors, available to the public only through a Freedom of Information request. Researchers who independently assessed risperidone and found that it appeared to cause motor dysfunction just as frequently as haloperidol did (or even more frequently) didn't have the finances to hire PR firms to publicize their research. Their papers quietly appeared in the medical journals, and the lay public never heard a peep about them. Even the criticism in *Lancet* didn't stir any bad newspaper press for Janssen. Besides, Eli Lilly had gained approval to market olanzapine in 1996, and that had spurred the press to burnish the atypicals story anew.

Play It Again, Sam

During the past fifteen years, most pharmaceutical research has focused on developing drugs that act narrowly on targeted receptors, with the thought that such "clean" drugs will have fewer side effects. Olanzapine, while a blockbuster financial success, ironically took antipsychotic drug development in the opposite direction. It, like clozapine, is a "dirty" drug. It acts on a broad range of receptors—dopaminergic, serotonergic, adrenergic, cholinergic, and histaminergic—and blocking any one of those receptors is known to cause an array of side effects. Blocking dopaminergic receptors leads to motor dysfunction. Blocking serotonergic receptors leads to sexual dysfunction, hypotension, and weight gain. Drugs that act on adrenergic receptors may cause hypotension, dizziness, tachycardia, and ejaculatory dysfunction. Anticholinergics may cause blurred vision, dry mouth, constipation, urinary retention, memory problems, drowsiness, fatigue, and erectile dysfunction. Blockade of histaminergic receptors can cause sedation and weight gain. The laundry list of possible side effects from a "dirty" drug like olanzapine is a long one. How this blockade of multiple receptors will play out in the human brain is also anybody's guess. It's a scientific crapshoot, but in the tale told to the public, this "dirty" aspect of olanzapine was transformed into a virtue. "Olanzapine," the Associated Press reported, might be better than risperidone "because it appears to affect even more areas of the brain."[37]

As was the case with risperidone, the FDA's review of the trial data for olanzapine revealed just how far Eli Lilly had spun the trial results.[38] First, Leber and another FDA official, Paul Andreason, found that Eli Lilly's studies were "biased against haloperidol" in much the same way that Janssen's had been. Multiple doses of olanzapine were compared to one dose of haloperidol, and the drugs were not compared at "equieffective" doses. In addition, many of the patients in the trials had previously taken haloperidol and presumably had not responded well to it, and including such "bad responders," the FDA officials noted, made it likely that results for haloperidol would be worse than normal, and thus help make olanzapine look superior by comparison. Concluded Leber: "The sample of patients used is an inappropriate choice" for comparison purposes. Second, he and Andreason determined that of Eli Lilly's four well-controlled studies, only the smaller two—with a combined total of about 500 patients—provided any useful data related to olanzapine's effectiveness versus placebo. In one of its two larger trials, involving 431 patients, Eli Lilly had compared three doses of olanzapine to haloperidol and to a low, nontherapeutic dose of olanzapine (which served as a placebo control), and Leber and Andreason concluded it was a "failed" study because there was no significant difference in the reduction of positive symptoms in any of the treatment groups at the end of six weeks. The other large study that the FDA found wanting was Eli Lilly's large phase III trial, involving 1,996 patients. This was the study that Eli Lilly had used to make claims in the medical journals that olanzapine was superior to haloperidol, and also the one that led to newspaper stories about how olanzapine was a "potential breakthrough of tremendous magnitude." However, both Leber and Andreason concluded that the study was "biased against haloperidol," and they detailed specific methods that Eli Lilly had used to favor its drug. Furthermore, since the study didn't include a placebo arm, it couldn't show any efficacy data in that regard, either. Concluded Andreason: The study "is fundamentally flawed and provides little useful efficacy data."

Olanzapine's safety profile was also not as benign as the newspaper reports suggested. Of the 2,500 patients in the trials who received olanzapine, twenty died. Twelve killed themselves, and two of the remaining eight deaths, both from "aspiration pneumonia," were seen by FDA reviewers as possibly causally related to olanzapine. Twenty-two percent of the olanzapine patients suffered a "serious" adverse event, compared to 18 percent of the haloperidol patients. Two-thirds of the olanzapine patients didn't successfully complete the trials. More than one-fourth of the patients complained that the drug made them sleepy. Weight gain was a frequent problem, with olanzapine patients putting on nearly a pound a week in the short-term trials, and twenty-six pounds over the course of a year (for those who participated in the extension trials).[39] Other problems that showed up, with greater or lesser frequency, included Parkinson's, akathisia, dystonia, hypotension, constipation, tachycardia, diabetic complications, seizures, increases in serum prolactin (which may cause leaking breasts and impotence and which raises the risk of breast cancer), liver abnormalities, and both leukopenia and neutropenia (white blood cell disorders). Leber, in his summation of the safety data, even warned that, given olanzapine's broad action on multiple receptor types, "no one should be surprised if, upon marketing, events of all kinds and severity not previously identified are reported in association with olanzapine's use."

The third atypical to undergo the FDA's review was AstraZeneca's quetiapine, and once again, the FDA found plenty to criticize.[40] Four of the eight trials conducted by AstraZeneca were not considered by the FDA to provide any "meaningful" efficacy data. The other four studies, the FDA determined, showed that quetiapine was modestly superior to placebo for reducing positive symptoms but did not prove that quetiapine was superior to haloperidol in this regard. If anything, trial data suggested that haloperidol was more effective. Patients also clearly had difficulty staying on quetiapine. Eighty percent of the 2,162 quetiapine-treated patients dropped out of the trials, compared to 61 percent of the placebo patients and 42 percent of patients treated with standard neuroleptics. Common adverse events included weight gain, sedation, and somnolence; there were also reports of hypotension, tachycardia, seizures, leukopenia, neutropenia, neuroleptic malignant syndrome, liver abnormalities, and bone fractures caused by fainting spells.

Three atypicals reviewed by the FDA, and three times the FDA did not find any convincing evidence that they were superior to the old ones. Instead, FDA reviewers pointed out the ways in which Janssen and Eli Lilly had used biased trial designs to produce results that, when published in the science journals, created a story of superiority (and enabled them to sell their new drugs for ten to thirty times the price of the old neuroleptics). However, such criticism did not require the knowledge of an FDA expert. The methods used by drug companies to make their drugs look good in clinical trials have become so well known that various articles have appeared in medical journals cataloging them. The use of inappropriate doses of the standard drug is a favorite one; so is comparing multiple dosages of the experimental drug to one dose of the standard drug. Yet when the researchers who'd been paid by Janssen and Eli Lilly to conduct the trials reported their results (or put their names on papers written by the companies), they never discussed how the trials were biased by design. They never fessed up, as it were, and their silence spoke volumes about the influence of money.[41]

Every once in a while, a researcher has stepped forward to poke holes in the atypicals story. A team of English scientists, led by John Geddes at the University of Oxford, reviewed results from fifty-two studies, involving 12,649 patients, and concluded in 2000, "there is no clear evidence that atypical antipsychotics are more effective or are better tolerated than conventional antipsychotics." The most common ruse that had been employed to make the drugs look better, Geddes found, was the use of "excessive doses of the comparator drug."[42] An embarrassing, yet revealing, squabble also briefly erupted in the medical journals over the relative merits of risperidone and olanzapine, with Janssen complaining that Eli Lilly's studies were biased in ways that—surprise, surprise—favored olanzapine. Then Janssen funded a comparative trial, and that trial concluded risperidone was superior to olanzapine. It all made for a tawdry spectacle, and finally a truce of sorts was called. Several studies concluded that it was impossible to say one or the other was better; they were different drugs, with different risk-benefit profiles, and perhaps it was best to leave it at that.[43]

Indeed, why would either company want to stir the pot? Both risperidone and olanzapine had quickly become astonishing financial successes. Total annual U.S. sales of antipsychotic medications roared past $2.5 billion in 2000. Worldwide sales of olanzapine were projected to hit $3 billion in 2001. As Forbes.com crowed on January 25, 2000: "Zyprexa (olanzapine) and its main competitor, Risperdal, can split a lot of market between them. Since they are both expensive drugs, they will fill company coffers." Praise from newspaper and magazine writers continued to flow as well. American science, the *Washington Post* told its readers on July 29, 1998, had developed several "breakthrough" medications that "have proven to be much more effective than older medications in helping schizophrenics lead functional lives and with far fewer side effects."[44] Money, glowing press—this was a good-news story all around, and finally the National Alliance for the Mentally Ill put it together into its full mythic glory. In 1999, it published a book titled *Breakthroughs in Antipsychotic Medications* and inside the front cover were framed, color photos of the new wonder pills. The NAMI authors wrote: "Conventional antipsychotics all do about the same job in the brain. They all correct brain chemistry by working on the dopamine systems in the brain . . . the newer medications seem to do a better job of balancing all of the brain chemicals, including dopamine and serotonin . . . give the new medication plenty of time to do a good job!"[45]

Like tonics once pitched from the backs of wooden wagons, atypicals could apparently transform the troubled mind into one awash with chemicals operating in perfect harmony.*

A State of Confusion

One of the saddest aspects of this "research" process, and the storytelling that accompanied it, is how it has left everyone in the dark about the real merits of the atypicals. There are certainly many

*Much like the academic doctors, NAMI is also the recipient of drug money. From 1996 to 1999, drug companies gave NAMI $11.72 million for a "Campaign to End Discrimination" against the mentally ill. The two largest donors were Eli Lilly ($2.87 million) and Janssen ($2.08 million). In addition, an Eli Lilly executive was "loaned" to NAMI in 1999 and helped the advocacy group with its "strategic planning."

anecdotal accounts of patients who are doing well on them, and so perhaps in some ways they truly are superior to the old drugs. Yet anecdotes do not make for good science, and the testing process was such that little can be known for certain. Are the atypicals, for instance, even any better than placebo at knocking down psychosis? If patients suffering a first episode of psychosis were separated into two groups and one group were given a placebo and the other olanzapine, what would the results be at the end of six weeks? Or perhaps more to the point, if a sedative were compared to olanzapine or risperidone, what would be the results? No one knows. As for their comparative merits versus standard neuroleptics—again, who knows? In fact, it is actually quite easy to envision a scenario in which haloperidol would be the drug being hailed today as the new wonder medication and olanzapine would be the drug being carted off to the trash heap. All one has to do is imagine their coming to market in reverse order, such that in 1995 olanzapine had been the "old" drug and haloperidol the "experimental" drug. In that case, multiple doses of haloperidol would have been compared to a single, high dose of olanzapine—in other words, the trials would have been designed to favor haloperidol—and researchers would likely have been able to announce that haloperidol appeared superior in several ways and didn't cause the troublesome side effects associated with olanzapine, like weight gain and sleepiness. The researchers would even have been able to offer a good explanation for why haloperidol had a superior side-effect profile. Whereas olanzapine was a "dirty" drug that acted on multiple neurotransmitters, haloperidol was a clean drug that more precisely honed in on a very specific receptor, the D_2 receptor. Modern science had simply produced a more refined drug.

The biggest question, of course, is how the new drugs will affect patients' lives over longer periods of time. The old drugs—as was shown by the WHO studies—led to an increase in chronic illness and limited the possibility of recovery. They were harmful over the long run. Will the new drugs be equally harmful? Less harmful? Or, in fact, helpful over the long term? No one knows. However, there are already plenty of reasons to worry about their long-term effects. The atypicals—just like standard neuroleptics—cause an abnormal increase in D_2 receptors.[46] And while certain side effects, such as the risk of tardive dyskinesia, may be reduced with the atypicals, they also bring their own set of new problems. For instance, there have been reports that olanzapine can trigger obsessive compulsive disorder, with researchers speculating that this may be due to the drug's hindrance of serotonin activity. Then there are the metabolic problems associated with olanzapine: Just how great is the increased risk of poor health with this drug because of weight gain? Some patients are putting on sixty, seventy, eighty pounds. Reports are also filtering into the medical literature about how olanzapine can dramatically increase triglyceride and blood sugar levels, which are risk factors for cardiovascular disease and diabetes. Is this a drug that will lead to early death for many?

What makes this question all the more pressing is that there remains today great uncertainty over what schizophrenia is, or isn't. The public has been led to think of schizophrenia as a discrete disorder, one characterized by abnormal brain chemistry. In truth, the biological underpinnings of madness remain as mysterious as ever. In fact, schizophrenia is a diagnosis applied to people who behave or think strangely in a variety of *different* ways. Some people so diagnosed are withdrawn. Some are manic. Some act very "silly." Others are paranoid. In some people, the crazy behaviors appear gradually. In others, psychosis descends abruptly. Any well-reasoned concept of "madness" would require teasing apart all these different types and would also require an understanding of how outcomes for the different types—in the absence of neuroleptics—might differ. Yet there is little research in American circles devoted to seeing this more complex picture. It is a shortcoming so pronounced that it caused Nancy Andreasen, editor of the *American Journal of Pyschiatry*, to burst forth in 1998 with a remarkable confession: "Someday in the twenty-first century, after the human genome and the human brain have been mapped, someone may need to organize a reverse Marshall Plan so that the Europeans can save American science by helping us figure out who really has schizophrenia or what schizophrenia really is."[47]

Two hundred years after Benjamin Rush founded American psychiatry, and still the problem remains as confounding as ever. What is madness? Where do you draw the line separating the normal mind from the crazy one? The drawing of that line is a profound event for a society, and a life-altering event for those diagnosed as ill. And it is here that one can see, once again, how the storytelling that brought the atypicals to market is exacting a great cost. With the new drugs presented to the public as wonderfully safe, American psychiatrists are inviting an ever-greater number of patients into the madness tent. They are prescribing atypicals for a wide range of emotional and behavioral disorders, and even for disruptive children, including—as the *Miami Herald* reported—toddlers only two years old. Yale University psychiatrists are even giving olanzapine to teenagers who are not even ill but simply said to be at risk of developing schizophrenia, either because they have siblings diagnosed with the disorder or have begun behaving in troubling ways.* Researchers, the *Wall Street Journal* reported, "hope the new drugs will intervene in the brain-damaging process that leads to schizophrenia, even though they don't know for sure what that process is."[48]

That is the story in American mad medicine today: The line between the sane and the not-so-sane is now being drawn in such a way that two-year-olds can be put on "antipsychotic" medications, and some researchers are busily speculating that their wonderful new drugs can stop an unknown brain-damaging process in people who aren't yet ill. Madness is clearly afoot in American psychiatry, and bad science—as so often has been the case in mad medicine—has helped it on its way.

*It also didn't take long for documents to surface suggesting that the teenagers recruited into the Yale study, and their families, were being misled. On December 12, 2000, the federal Office for Human Research Protections criticized the Yale investigators for using informed consent forms that "failed to include an adequate description of the reasonably foreseeable risks and discomforts." In the consent forms, the Yale researchers had told the young adults, who were not ill, that "while the clinical goal is to help you feel better and in more control of your life, it is possible that you will feel worse. This is a risk of your clinical condition, not a risk of being in the study." Such wording, the OHRP noted, did not "take into account 'feeling worse' due to olanzapine side effects."

EPILOGUE

Biological psychiatry, as always, promises us that a medical solution is almost within our grasp. It would be nice if one could believe it. I fear one might as well be waiting for Godot.

—Andrew Scull[1]

THIS BOOK BEGAN with a straightforward goal, and that was to explore why schizophrenia outcomes are so poor in the United States today. It seemed like a simple question, and yet it quickly opened the door to a larger story—the story of how we as a society have historically treated those we call "mad." It clearly is a troubled history, one that begs to be better known. There are, perhaps, many lessons that can be drawn from it, but one seems to stand out above all others. Any hope of reforming our care of those "ill with schizophrenia" will require us to rediscover, in our science, a capacity for humility and candor.

There is one moment in the past where we can find such humility. It can be seen in moral therapy as practiced in its most ideal form, by the Quakers in York, England, or by Thomas Kirkbride at the Pennsylvania Hospital for the Insane in the mid-nineteenth century. In their writings, the York Quakers regularly confessed that they understood little about any possible physical causes of madness. But what they did see clearly was "brethren" who were suffering and needed comfort. That was the understanding that drove their care, and so they sought to run their asylum in a way that was best for their patients, rather than in a way that was best for them, as managers of the asylum. They put their patients' comforts and needs *first.* They also perceived of their patients as having a God-given capacity for recovery, and thus simply tried to "assist Nature" in helping them heal. It was care that was at once humanitarian and optimistic, and it did help many get well. But equally important, the York Quakers were quite willing to accept that many of their brethren would continue in their crazy ways. That was all right, too. They would provide a refuge for those who could not regain their mental health and at least make sure they had warm shelter and good food.

In the 1960s, as the United States set out to reform its care, it did look back to moral treatment for inspiration. President John Kennedy and the Joint Commission on Mental Illness and Mental Health spoke of the need for American society to see those who were distraught in mind as part of the human family, and deserving of empathy. Eugenics had stirred America to treat the severely mentally ill with scorn and neglect, and it was time to change our ways. We would welcome the mentally ill back into society. Asylums would be replaced with community care. But the design of that reform also rested on a medical notion of the most unusual sort, that neuroleptics "might be described as moral treatment in pill form." The confusion in that perception was profound: Neuroleptics were a medical treatment with roots in frontal lobotomy and the brain-damaging therapeutics of the eugenics era. Our vision for reform and the medical treatment that would be the cornerstone of that reform were hopelessly at odds.

Something had to give, and the moment of choice occurred very early on. The research study that launched the emptying of the state hospitals was the six-week trial conducted by the National Institute of Mental Health in the early 1960s, which concluded that neuroleptics were safe and antischizophrenic. But then, a very short while later, the NIMH found in a follow-up study that the patients who had been treated with neuroleptics were more likely than the placebo patients to have been rehospitalized. Something clearly was amiss. A choice, in essence, was presented to psychiatry. Would it hold to the original vision of reform, which called for the provision of care that would promote *recovery*? If so, it would clearly need to rethink the merits of neuroleptics. The drugs were apparently making people chronically ill, and that was quite apart from whatever other drawbacks they might have. Or would it cast aside questions of recovery and instead defend the drugs?

There can be no doubt today about which choice American psychiatry made. Evidence of the harm caused by the drugs was simply allowed to pile up and up, then pushed away in the corner where it wouldn't be seen. There was Bockoven's study that relapse rates were lower in the pre-neuroleptic era. Rappaport's study. Mosher's. Reports of neuroleptic malignant syndrome and tardive dyskinesia. Van Putten's report of medicated patients in boarding homes spending their days idly looking at television, too numbed in mind and spirit to even have a favorite program. Studies detailing the high incidence of akathisia, Parkinson's, and a myriad of other types of motor dysfunction. Case reports of akathisia driving patients so out of their minds it made them suicidal or even homicidal. Harding's study and then the WHO studies. All of this research told of suffering, and of loss. And where were the studies showing that the drugs were leading people to *recovery*? Researchers studiously avoided this question. In 1998, British investigators reviewed the published results of 2,000 clinical trials of neuroleptics over the previous fifty years and found that only one in twenty-five studies even bothered to assess "daily living activities" or "social functioning."[2] The trials again and again simply looked at whether the drugs knocked down visible symptoms of psychosis and ignored what was really happening to the patients as *people*.

It is not difficult today to put together a wish list for reform. An obvious place to start would be to revisit the work of Emil Kraepelin. Were many of his psychotic patients actually suffering from encephalitis lethargica, and has that led to an overly pessimistic view of schizophrenia? The next step would be to investigate what the poor countries are doing right. How are the "mad" treated in India and Nigeria? What are the secrets of care—beyond not keeping patients regularly medicated—that help so many people in those countries get well? Closer to home, any number of studies would be welcome. A study that compares neuroleptics to sedatives would be helpful. How would conventional treatment stack up against care that provided "delusional" people with a safe place to live, food, and the use of sedatives to help restore their sleep-wake cycles? Or how about an NIMH-funded experiment modeled on the work of Finnish investigators? There, physicians led by Yrjö Alanen at the University of Turku have developed a treatment program that combines social support, family therapy, vocational therapy, and the selective use of antipsychotics. They are picking apart differences in patient types and have found that some patients do better with low doses of antipsychotics, and others with no drugs at all. They are reporting great results—a majority of patients so treated are remaining well for years, and holding jobs—so why not try it here?

At the top of this wish list, though, would be a simple plea for honesty. Stop telling those diagnosed with schizophrenia that they suffer from too much dopamine or serotonin activity and that the drugs put these brain chemicals back into "balance." That whole spiel is a form of medical fraud, and it is impossible to imagine any other group of patients—ill, say, with cancer or cardiovascular disease—being deceived in this way.

In truth, the prevailing view in American psychiatry today is that there are any number of factors—biological and environmental—that can lead to schizophrenia. A person's genetic makeup obviously may play a role. Relatives of people with schizophrenia appear to be at increased risk of developing the disorder, and thus the thought is that they may inherit genes that make them less able to cope with environmental stresses. The genetic factors are said to *predispose* people to schizophrenia, rather than cause it. Another prominent theory is that complications during pregnancy or during delivery may affect the developing brain, and that this trauma leads to deficiencies in brain function once neuronal systems have matured. Yet another thought is that some people with schizophrenia have difficulty filtering incoming sensory data, and that this problem is due to abnormal function in brain cells known as interneurons. A number of investigators are still studying the role that different neurotransmitters may play in the disorder. The biological paths to schizophrenia may be many, but none is yet known for sure. It is also possible that the capacity to go mad, as it were, is in all of us. Extreme emotional trauma can clearly trigger psychosis, and some argue that psychosis is a mechanism for coping with that trauma. That view of the disorder is consistent with the fact that in the absence of neuroleptics, many people who suffer a schizophrenic break recover from it, and never relapse again.

Thus, if we wanted to be candid today in our talk about schizophrenia, we would admit to this: Little is known about what causes schizophrenia. Antipsychotic drugs do not fix any known brain abnormality, nor do they put brain chemistry back into balance. What they do is alter brain function in a manner that diminishes certain characteristic symptoms. We also know that they cause an increase in dopamine receptors, which is a change associated both with tardive dyskinesia and an increased biological vulnerability to psychosis, and that long-term outcomes are much better in countries where such medications are less frequently used. Although such candor might be humbling to our sense of medical prowess, it might also lead us to rethink what we, as a society, should do to help those who struggle with "madness."

But, none of this, I'm afraid, is going to happen. Olanzapine is now Eli Lilly's top-selling drug, surpassing even Prozac. There will be no rethinking of the merits of a form of care that is bringing profits to so many. Indeed, it is hard to be optimistic that the future will bring any break with the past. There is no evidence of any budding humility in American psychiatry that might stir the introspection that would be a necessary first step toward reform. At least in the public arena, all we usually hear about are advancements in knowledge and treatment, as if the march of progress is certain. Eli Lilly and Janssen have even teamed up with leaders of U.S. mental-health advocacy groups to mount "educational" missions to poor countries in East Asia, so that we can export our model of care to them.[3] Hubris is everywhere, and in mad medicine, that has always been a prescription for disaster. In fact, if the past is any guide to the future, today we can be certain of only one thing: The day will come when people will look back at our current medicines for schizophrenia and the stories we tell to patients about their abnormal brain chemistry, and they will shake their heads in utter disbelief.

NOTES

Additional information on sources can be found at: www.madinamerica.com

Chapter I: Bedlam in Medicine

1. Benjamin Rush, *Medical Inquiries and Observations upon the Diseases of the Mind* (reprint of 1812 edition; Hanger Publishing, 1962), 211.

2–7. Thomas Morton, *The History of the Pennsylvania Hospital* (Times Printing House, 1895), 144, 8, 147, 163, 130, 148.

8. Rush, *Medical Inquiries*, 178.

9. Andrew Scull details the history of English physicians conceiving of the mad as wild animals in *Social Order/Mental Disorder* (University of California Press, 1989), 54–79.

10. As cited by Gregory Zilboorg, *A History of Medical Psychology* (W. W. Norton, 1941), 261.

11. As cited by Scull, *Social Order/Mental Disorder*, 58.

12. As cited by Richard Hunter and Ida Macalpine, *Three Hundred Years of Psychiatry* (Oxford University Press, 1963), 705.

13. As cited by Scull, *Social Order/Mental Disorder*, 62.

14. William Cullen, *Practice of Physic* (L. Riley, 1805), 489.

15. George Man Burrows, *Commentaries on the Causes, Forms, Symptoms, and Treatment, Moral and Medical, of Insanity* (reprint of 1828 edition; The Classics of Psychiatry and Behavioral Sciences Library, 1994), 640.

16. Ibid., 642–643.

17. Emil Kraepelin, *One Hundred Years of Psychiatry* (Philosophical Library, 1962), 61.

18. Cullen, *Practice of Physic*, 488.

19. Andrew Scull, *Masters of Bedlam: The Transformation of the Mad-Doctoring Trade* (Princeton University Press, 1996), 279.

20. Ida Macalpine and Richard Hunter, *George III and the Mad-Business* (Penguin Press, 1969), 323.

21. William Battie, *A Treatise on Madness* (reprint of 1758 edition; Brunner/Mazel, 1969), 93.

22–25. Macalpine and Hunter, *George III*, 47–86, 291, 328.

26. Burrows, *Commentaries on the Causes, Forms, Symptoms, and Treatment*, 528.

27. Cullen, *Practice of Physic*, 490.

28. As cited by Zilboorg, *History of Medical Psychology*, 298.

29. Burrows, *Commentaries on the Causes, Forms, Symptoms, and Treatment*, 627.

30. As cited by Scull, *Social Order/Mental Disorder*, 68.

31. As cited in ibid., 71.

32. Kraepelin, *One Hundred Years of Psychiatry*, 87–88.

33. Burrows, *Commentaries on the Causes, Forms, Symptoms, and Treatment*, 532, 606.

34. As cited by Scull, *Social Order/Mental Disorder*, 64.

35. Thomas Percival, *Medical Ethics* (reprint of 1803 edition; DevCom, 1987), 29.

36. As cited by O. H. Perry Pepper, "Benjamin Rush's Theories on Blood Letting After One Hundred and Fifty Years," *Transactions and Studies of the College of Physicians of Philadelphia* 14 (1946), 121–126.

37. Richard Shryock, *Medicine and Society in America: 1660–1860* (Cornell University Press, 1962), 1–38.

38. Rush, *Medical Inquiries*, 17.

39. Shryock, *Medicine and Society in America*, 31–32.

40. Rush, *Medical Inquiries*, 104, 175, 198, 212, 224.

41. Burrows, *Commentaries on the Causes, Forms, Symptoms, and Treatment*, 602.

42. As cited by Scull, *Social Order/Mental Disorder*, 69.

43. Kraepelin, *One Hundred Years of Psychiatry*, 17.

44. As cited by Norman Dain, *Concepts of Insanity in the United States, 1789–1865* (Rutgers University Press, 1964), 24.

45. As cited by Mary Ann Jimenez, *Changing Faces of Madness: Early American Attitudes and Treatment of the Insane* (University Press of New England, 1987), 110.

46. As cited by Albert Deutsch, *The Mentally Ill in America* (Doubleday, Doran and Company, 1937), 140.

Chapter 2: The Healing Hand of Kindness

1. As cited by Gerald Grob, *The Mad Among Us* (Harvard University Press, 1994), 66.

2. Philipe Pinel, *A Treatise on Insanity* (reprint of 1806 edition; Hafner Publishing, 1962), 32.

3. Ibid., 108.

4. Samuel Tuke, *Description of the Retreat* (reprint of 1813 edition; Process Press, 1996), 128.

5. As cited by Grob, *The Mad Among Us*, 30.

6. As cited by Scull, *Social Order/Mental Disorder*, 102.

7. As cited in ibid., 105.

8. *Harpers's Weekly*, March 19, 1859, 185; as cited by Lynn Gamwell and Nancy Tomes, *Madness in America* (Cornell University Press, 1995), 60.

9. In *Mental Institutions in America: Social Policy to 1875* (Free Press, 1973), Gerald Grob details outcomes for the early moral treatment asylums. The recovery rates listed for the individual asylums come from the following sources: (1)

Bloomingdale Asylum, *Mental Institutions in America,* 68; (2) Worcester State Lunatic Hospital, *Annual Reports* (1835, 1836, and 1841), as cited by Grob in *The Mad Among Us,* 99; (3) Hartford Retreat, *Third Annual Report* (1827), as cited by Scull, *Social Order/Mental Disorder,* 110; (4) McLean Hospital, as cited by Grob, *The Mad Among Us,* 36; and (5) Friend's Asylum, as cited by Dain, *Concepts of Insanity in the United States,* 120, 132.

10. As cited by Deutsch, *The Mentally Ill in America,* 151.

11. In *Social Order/Mental Disorder,* Andrew Scull details how the first corporate asylums in the United States were modeled after the York Retreat, and thus medicine played a secondary role.

12. Dain, *Concepts of Insanity in the United States,* 154.

13. As cited by Scull, *Social Order/Mental Disorder,* 103.

14. As cited in ibid., 106.

15. Deutsch, *The Mentally Ill in America,* 215.

16. Earle Pliny, "Bloodletting in Mental Disorders," *American Journal of Insanity* 10 (1854):397. Also see Nancy Tomes, *The Art of Asylum Keeping* (University of Pennsylvania Press, 1994), 74–89.

17. Edward Jarvis, "On the Supposed Increase of Insanity," *American Journal of Insanity* 8 (April 1852):333–364.

18. The principal source for Thomas Kirkbride's governance of Pennsylvania Hospital is Tomes, *The Art of Asylum Keeping.*

19. As cited by Gamwell and Tomes, *Madness in America,* 93.

20. Tomes, *The Art of Asylum Keeping,* 139.

21. Morton, *History of Pennsylvania Hospital,* 172.

22. Tomes, *The Art of Asylum Keeping,* 218.

23. Ibid., 224–226.

24. Dorothea Dix, *On Behalf of the Insane Poor: Selected Reports* (Arno Press, 1971), 4.

25. As cited by Grob, *The Mad Among Us,* 94.

26. J. Sanbourne Bockoven, *Moral Treatment in Community Mental Health* (Springer Publishing Company, 1972), 15.

27. For perspectives on the efficacy of moral treatment, see Scull in *Social Order/Mental Disorder,* 90; Grob in *The Mad Among Us,* 99–102; Dain in *Concepts of Insanity in the United States,* 120, 132; Morton, *History of Pennsylvania Hospital,* 243; and Bockoven in *Moral Treatment in Community Mental Health,* 14–15, 55–67.

28. Edward Spitzka, "Reform in the Scientific Study of Psychiatry," *Journal of Nervous and Mental Disease* 5 (1878):201–229; Edward Seguin, "Neurological Correspondence," *Journal of Nervous and Mental Disease* 5 (1878): 336; S. Weir Mitchell, "Address Before the Fiftieth Annual Meeting of the American Medico-Psychological Association," *Journal of Nervous and Mental Disease* 21 (1894): 413–437; William Hammond, "The Non-Asylum Treatment of the Insane," *Transactions of the Medical Society of New York* (1879):280–297. Also see Bonnie Blustein's essay in *Madhouses, Mad-Doctors, and Madmen,* ed. Andrew Scull (University of Pennsylvania Press, 1981), 241–270.

29. Edward Cowles, "The Advancement of Psychiatry in America," *American Journal of Insanity* 52 (1896):364–386.

Chapter 3: Unfit To Breed

1. Alexis Carrel, *Man the Unknown* (Harper and Brothers, 1935), 318.

2. As cited by Daniel Kevles, *In the Name of Eugenics* (University of California Press, 1985), 3. Biographical details on Galton are primarily from Kevles's book.

3. As cited by Allan Chase, *The Legacy of Malthus* (University of Illinois Press, 1980), 85, 101.

4. As cited by Peter Medawar, *Aristotle to Zoos* (Harvard University Press, 1983), 87.

5. As cited by Chase, *Legacy of Malthus*, 13.

6. As cited by Kevles, *In the Name of Eugenics*, 12.

7. As cited by Edward Shorter, *A History of Psychiatry* (John Wiley and Sons, 1997), 96.

8. Henry Maudsley, *The Pathology of the Mind* (reprint of 1895 edition; Julian Friedmann Publishers, 1979), 47.

9. Gerald Grob, *Mental Illness and American Society, 1875–1940* (Princeton University Press, 1983), 8.

10. As cited by Chase, *Legacy of Malthus*, 8.

11. As cited by Kevles, *In the Name of Eugenics*, 85.

12. As cited by Charles Rosenberg, *No Other Gods: On Science and American Social Thought* (Johns Hopkins University Press, 1976), 90.

13. Charles Davenport, *Heredity in Relation to Eugenics* (Henry Holt, 1911), 216–219.

14. Ibid., iv.

15. As cited by Kevles, *In the Name of Eugenics*, 53.

16. Charles Robinson, ed., *The Science of Eugenics* (W. R. Vansant, 1917), 97.

17. Prescott Hall, "Immigration Restriction and Eugenics," *Journal of Heredity* 10 (1919):126; Seth Humphrey, "The Menace of the Half Man," *Journal of Heredity* 11 (1920):231; Robert Sprague, "Education and Race Suicide," *Journal of Heredity* 6 (1915):158; and Paul Popenoe, "Harvard and Yale Birth Rates," *Journal of Heredity* 7 (1916):569. Popenoe quotes Phillips in his *Journal of Heredity* article.

18. Harry Laughlin, "The Progress of American Eugenics," *Eugenics* 2 (February 1929):3–16. Also *Eugenical News* 9 (December 1924):104.

19. As cited by Kevles, *In the Name of Eugenics*, 63.

20. Eugenics Record Office, Bulletin No. 10 B, "The Legal, Legislative, and Administrative Aspects of Sterilization," February 1914.

21. As cited by Chase, *Legacy of Malthus*, 15.

22. Race Betterment Foundation, *Proceedings of the First National Conference on Race Betterment*, Battle Creek, MI, January 8–12, 1914, 479.

23. A. J. Rosanoff, Eugenics Record Office, Bulletin No. 5, "A Study of Heredity of Insanity in the Light of the Mendelian Theory," October 1911.

24. Abraham Myerson, "A Critique of Proposed 'Ideal' Sterilization Legislation," *Archives of Neurology and Psychiatry* 33 (March 1935):453–463.

25. Robinson, *The Science of Eugenics*, 12.

26. Aaron Rosanoff, ed., *Manual of Psychiatry* (John Wiley and Sons, 1920). The quotation is from a review of the book in *Journal of Heredity* 12 (1921):300.

27. Paul Popenoe, "Heredity and the Mind," *Journal of Heredity* 7 (1916):456–462.

28. "His Trust in Eugenics Is Excessive," *New York Times* editorial, June 19, 1923.

29. Charles Gibbs, "Sex Development and Behavior in Male Patients with Dementia Praecox," *Archives of Neurology and Psychiatry* 9 (1923):73–87; Paul Popenoe, "Marriage Rates of the Psychotic," *Journal of Nervous and Mental Disease* 68 (1928):17–27; Paul Popenoe, "Fecundity of the Insane," *Journal of Heredity* 19 (1928):73–82; Sewall Wright, "Heredity and Mental Disease," *Journal of Heredity* 16 (1925):461–462; and Abraham Myerson, ed., *Eugenical Sterilization: A Reorientation of the Problem* (Macmillan, 1936).

30. Paul Popenoe and Roswell Johnson, *Applied Eugenics* (Macmillan, 1933), 134; and E. S. Gosney and Paul Popenoe, *Sterilization for Human Betterment* (Macmillan, 1929), 7.

31. See Barry Mehler, *A History of the American Eugenics Society, 1921–1940*, Ph.D. diss., University of Illinois at Urbana-Champaign, 1988, 36–60, for a description of the Second International Congress of Eugenics; also Chase, *Legacy of Malthus*, 277–284.

32. "Eugenics as Romance," *New York Times* editorial, September 25, 1921.

33. As cited by Mehler, *History of the American Eugenics Society*, 61.

34. Ibid, 129–179.

35. *Eugenical News* 10 (July 1925):131.

36. As cited by Kevles, *In the Name of Eugenics*, 62.

37. *Eugenical News* 10 (March 1925):27.

38. *Eugenics* 2 (August 1929):3–19; also see Mehler, *History of the American Eugenics Society*, 90.

39. As cited by Mehler, *History of the American Eugenics Society*, 246.

40. Frank Kallmann, "Heredity, Reproduction, and Eugenic Procedure in the Field of Schizophrenia," *Eugenical News* 23 (November–December 1938):105.

41. As cited by Mehler, *History of the American Eugenics Society*, 244.

42. Earnest Hooton, *Apes, Men, and Morons* (G. Putnam's and Sons, 1937), 269, 295.

43. Popenoe and Johnson, *Applied Eugenics*, 186.

44. Eugenics Record Office, Bulletin No. 10 B, "The Legal, Legislative, and Administrative Aspects of Sterilization," February 1914, 142.

45. Leon Cole, "Biological Eugenics," *Journal of Heredity* 5 (1914):305–312.

46. Myerson, *Eugenical Sterilization*, 24.

47. Paul Popenoe, "In the Melting Pot," *Journal of Heredity* 14 (1923):223.

48. William Goodell, "Clinical Notes on the Extirpation of the Ovaries for Insanity," *American Journal of Insanity* 38 (1882):293–302.

49. Cited by Angela Gugliotta, "Dr. Sharp with His Little Knife," *Journal of the History of Medicine* 53 (1998):371–406.

50. As cited by Kevles, *In the Name of Eugenics*, 93.

51. As cited by Julius Paul, essay in *Eugenic Sterilization*, ed. Jonas Robitscher (Charles C. Thomas, 1973), 25–40.

52. *Buck v. Bell*, 274 US 205 (1927). Also see Chase, *Legacy of Malthus*, 315–316.

53. As cited by Joel Braslow, *Mental Ills and Bodily Cures* (University of California Press, 1997), 56.

54. See Joel Braslow, "In the Name of Therapeutics: The Practice of Sterilization in a California State Hospital," *Journal of the History of Medicine* 51 (1996):29–51.

55. Ibid, 38, 44.

56. Gosney and Popenoe, *Sterilization for Human Betterment*, xiv, 33.

57. Popenoe, "Public Opinion on Sterilization in California," *Eugenical News* 20 (September 1935):73.

58. Gosney and Popenoe, *Sterilization for Human Betterment*, 32.

59. See Paul Weindling's essay, "The Rockefeller Foundation and German Biomedical Sciences, 1920–1940," in *Science, Politics, and the Public Good*, ed. Nicolaas Rupke (MacMillan Press, 1988), 119–140.

60. Margaret Smyth, "Psychiatric History and Development in California," *American Journal of Psychiatry* 94 (1938):1223–1236.

61. "Sterilization and Its Possible Accomplishments," *New England Journal of Medicine* 211 (1934):379–380; *New York Times* editorial, "Purifying the German Race," August 8, 1933; William Peter, "Germany's Sterilization Program," *American Journal of Public Health* 24 (March 1934):187–191; Andre Sofair, "Eugenic Sterilization and a Qualified Nazi Analogy: The United States and Germany, 1930–1945," *Annals of Internal Medicine* 132 (2000):312–319.

62. As cited by Kevles, *In the Name of Eugenics*, 116; and Sofair, "Eugenical Sterilization and a Qualified Nazi Analogy."

63. Davenport, *Heredity in Relation to Eugenics*, 263.

64. Madison Grant, *The Passing of the Great Race* (Charles Scribner's Sons, 1916), 45.

65. Stefan Kühl, *The Nazi Connection: Eugenics, American Racism, and German National Socialism* (Oxford University Press, 1994), 85.

66. "Exhibited as Case for Merciful Extinction," *New York Times*, February 7, 1921.

67. Hooton, *Apes, Men, and Morons*, 236, 294–295.

68. Carrel, *Man the Unknown*, 318–319.

69. Albert Deutsch, *The Shame of the States* (Harcourt, Brace, 1948), 41–42.

70. Albert Maisel, "Bedlam 1946," *Life*, May 6, 1946.

71. See Grob, *Mental Illness and American Society*, 190.

72. Ibid., 194.

73. Deutsch, *The Shame of the States*, 96.

74. Group for the Advancement of Psychiatry, Report No. 5 (April 1948), 1–19.

75. Marle Woodson, *Behind the Door of Delusion* (Macmillan Co., 1932; reprint edition, University Press of Colorado, 1994), 93.

76. John Maurice Grimes, *Institutional Care of Mental Patients in the United States* (self-published, 1934), xiv, 15–43, 95–99.

77. Harold Maine, *If a Man Be Mad* (Doubleday, 1947; republished by Permabooks, 1952), 309.

78. Deutsch, *The Shame of the States*, 57–58.

Chapter 4: Too Much Intelligence

1. Abraham Myerson in discussion of a paper by Franklin Ebaugh, "Fatalities Following Electric Convulsive Therapy," *Transactions of the American Neurological Association* 68 (1942):36–41.

2. William Russell, "From Asylum to Hospital: A Transition Period," *American Journal of Psychiatry* 100.2 (1944):87–97.

3. Edward Strecker, "The Continuous Bath in Mental Disease," *Journal of American Medical Association* 68 (1917):1796–1798; George Tuttle, "Hydrotherapeutics," *American Journal of Insanity* 61 (1904):179–192; G. W. Foster, "Common Features in Neurasthenia and Insanity," *American Journal of Insanity* 56 (1900):395–416.

4. Emmet Dent, "Hydriatric Procedures as an Adjunct in the Treatment of Insanity," *American Journal of Insanity* 59 (1902):91–100.

5. As cited by Braslow, *Mental Ills and Bodily Cures*, 50.

6. Herman Adler, "Indications for Wet Packs in Psychiatric Cases," *Boston Medical and Surgical Journal* 175 (1916):673–675.

7. As cited by Braslow, *Mental Ills and Bodily Cures*, 47.

8. J. Allen Jackson, "Hydrotherapy in the Treatment of Mental Diseases," *Journal of American Medical Association* 64 (1915):1650–1651.

9. W. O. Henry, "To What Extent Can the Gynecologist Prevent and Cure Insanity in Women," *Journal of American Medical Association* 48 (1907):997–1003.

10. Dr. Stone, "Proceedings of the Association," *American Journal of Insanity* 48 (1892):245–247.

11. Clara Barrus, "Gynecological Disorders and Their Relation to Insanity," *American Journal of Insanity* 51 (1895):475–489.

12. In his review of patient records at Stockton State Hospital, Braslow found five instances of clitoridectomy performed from 1947 to 1950 (Braslow, *Mental Ills and Bodily Cures*, 166).

13. William Mabon, "Thyroid Extract—A Review of the Results Obtained in the Treatment of One Thousand Thirty-Two Collected Cases of Insanity," *American Journal of Insanity* 56 (1899):257–273.

14. Leland Hinsie, "The Treatment of Schizophrenia: A Survey of the Literature," *Psychiatric Quarterly* 3 (1929):5–34; G. de M. Rudolf, "Experimental Treatments of Schizophrenia," *Journal of Mental Science* 77 (1931):767–791.

15. Henry Cotton, "The Relation of Oral Infection to Mental Diseases," *Journal of Dental Research* 1 (1919):269–313. Also see Andrew Scull, "Desperate Remedies: A Gothic Tale of Madness and Modern Medicine," *Psychological Medicine* 17 (1987):561–577.

16. Henry Cotton, "The Relation of Chronic Sepsis to the So-Called Functional Mental Disorders," *Journal of Mental Science* 69 (1923):434–462.

17. Quotation as cited by Scull, "Desperate Remedies." Relapse rate is from Henry Cotton, "The Etiology and Treatment of the So-Called Functional Psychoses," *American Journal of Insanity* 79 (1922):157–194.

18. As cited by Scull, "Desperate Remedies"; Cotton, "The Etiology and Treatment of the So-Called Functional Psychoses."

19. Remark made in discussion period following Cotton's 1922 paper, "The Etiology and Treatment of the So-Called Functional Psychoses."

20. As cited by Scull, "Desperate Remedies."

21. See Shorter, *History of Psychiatry*, 192–196, 200–207, for information on Klaesi's deep-sleep treatment and Julius Wagner-Jauregg's malaria therapy.

22. Hinsie, "The Treatment of Schizophrenia: A Survey of the Literature," 5–34; Leland Hinsie, "Malaria Treatment of Schizophrenia," *Psychiatric Quarterly* 1 (1927):210–214.

23. John Talbott, "The Effects of Cold on Mental Disorders," *Diseases of the Nervous System* 2 (1941):116–126; Douglas Goldman, "Studies on the Use of Refrigeration Therapy in Mental Disease with Report of Sixteen Cases," *Journal of Nervous and Mental Disease* 97 (1943):152–165.

24. See Grob, *Mental Illness and American Society*, 193–196.

25. Manfred Sakel, "The Origin and Nature of the Hypoglycemic Therapy of the Psychoses," *Bulletin of the New York Academy of Medicine* 13 (1937):97–109; Sakel, "A New Treatment of Schizophrenia," *American Journal of Psychiatry* 93 (1937):829–841; and Sakel, "The Methodical Use of Hypoglycemia in the Treatment of Psychoses," *American Journal of Psychiatry* 94 (1937):111–129.

26. Lothar Kalinowsky and Paul Hoch, *Shock Treatments and Other Somatic Procedures in Psychiatry* (Grune and Stratton, 1950), 82.

27. Manfred Sakel, *Schizophrenia* (Philosophical Library, 1958), 199, 261, 334.

28. Joseph Wortis, "Sakel's Hypoglycemic Insulin Treatment of Psychoses: History and Present Status," *Journal of Nervous and Mental Disease* 85 (1937):581–590; Wortis, "Further Experiences at Bellevue Hospital with the Hypoglycemic Insulin Treatment of Schizophrenia," *American Journal of Psychiatry* 94 (1937):153–158; Benjamin Malzberg, "Outcome of Insulin Treatment of One Thousand Patients with Dementia Praecox," *Psychiatric Quarterly* 12 (1938):528–553; John Ross, "A Review of the Results of the Pharmacological Shock Therapy and the Metrazol Convulsive Therapy in New York State, *American Journal of Psychiatry* 96 (1939):297–316.

29. "Mind Is Mapped in Cure of Insane," *New York Times*, May 15, 1937; "The Attack on Brainstorms," *Harper's*, 183 (1941):366–376; "Death for Sanity," *Time*, November 20 (1939):39–40; "Bedside Miracle," *Reader's Digest* 35 (1939):73–75.

30. Alexander Gralnick, "Psychotherapeutic and Interpersonal Aspects of Insulin Treatment," *Psychiatric Quarterly* 18 (1944):179.

31. Sakel, *Schizophrenia*, 261.

32. F. Humbert, "Critique and Indications of Treatments in Schizophrenia," *American Journal of Psychiatry*, supplement, 94 (1938):174–183.

33. Sakel, *Schizophrenia*, 319.

34. Joseph Wortis, "Case Illustrating the Treatment of Schizophrenia by Insulin Shock," *Journal of Nervous and Mental Disease* 85 (1937):446–456.

35. Ch. Palisa, "The Awakening from the Hypoglycemic Shock," *American Journal of Psychiatry*, supplement, 94 (1938):96–108.

36. Marcus Schatner, "Some Observations in the Treatment of Dementia Praecox with Hypoglycemia," *Psychiatric Quarterly* 12 (1938):5–29.

37. Palisa, "Awakening from Hypoglycemic Shock."

38. Kalinowsky and Hoch, *Shock Treatments and Other Somatic Procedures*, 69–70; Sakel, *Schizophrenia*, 331; Palisa, "Awakening from Hypoglycemic Shock," 102.

39. Solomon Katzenelbogen, "A Critical Appraisal of the 'Shock Therapies' in the Major Psychoses, II-Insulin," *Psychiatry* 3 (1940):211–228; Nolan Lewis, "The Present Status of Shock Therapy of Mental Disorders," *Bulletin of the New York Academy of Medicine* 19 (1943):227–243; Sakel, *Schizophrenia*, 238; Kalinowsky and Hoch, *Shock Treatments and Other Somatic Procedures*, 81–83; Humbert, "Critique and Indications of Treatments in Schizophrenia"; and Edwin Kepler, "The Psychiatric Manifestations of Hypoglycemia," *American Journal of Psychiatry* 94 (1937):89–108.

40. Marie Beynon Ray, *Doctors of the Mind* (Little, Brown and Company, 1946), 242.

41. D. M. Palmer, "Insulin Shock Therapy, a Statistical Survey of 393 Cases," *American Journal of Psychiatry* 106 (1950):918–925; Leon Salzman, "An Evaluation of Shock Therapy," *American Journal of Psychiatry* 103 (1947):669–679; Gralnick, "Psychotherapeutic and Interpersonal Aspects of Insulin Treatment."

42. Harold Bourne, "The Insulin Myth," *Lancet* 2 (1953):964–968.

43. Joseph Wortis, "On the Response of Schizophrenic Subjects to Hypoglycemic Insulin Shock," *Journal of Nervous and Mental Disease* 85 (1936):497–505; Malzberg, "Outcome of Insulin Treatment of One Thousand Patients with Dementia Praecox."

44. William Ruffin, "Attitudes of Auxiliary Personnel Administering Electroconvulsive and Insulin Coma Treatment: A Comparative Study," *Journal of Nervous and Mental Disease* 131 (1960):241–246; David Wilfred Abse, "Transference and Countertransference in Somatic Therapies," *Journal of Nervous and Mental Disease* 123 (1956):32–39.

45. Max Fink, "Meduna and the Origins of Convulsive Therapy," *American Journal of Psychiatry* 141 (1984):1034–1041.

46. Ladislaus von Meduna, "General Discussion of the Cardiazol Therapy," *American Journal of Psychiatry,* supplement, 94 (1938):41–50.

47. Horatio Pollock, "A Statistical Study of 1,140 Dementia Praecox Patients Treated with Metrazol," *Psychiatric Quarterly* 13 (1939):558–568.

48. Abram Elting Bennett, *Fifty Years in Neurology and Psychiatry* (Intercontinental Medical Book Corporation) (1972), 92; Broughton Barry, "The Use of Cardiazol in Psychiatry," *Medical Journal of Australia* (1939); as cited by Leonard Frank, *The History of Shock Treatment* (self-published, 1978), 12.

49. Lawrence Geeslin, "Anomalies and Dangers in the Metrazol Therapy of Schizophrenia," *American Journal of Psychiatry* 96 (1939):183; Polloack, "A Statistical Study of 1,140 Dementia Praecox Patients Treated with Metrazol"; Simon Kwalwasser, "Report on 441 Cases Treated with Metrazol," *Psychiatric Quarterly* 14 (1940):527–546; Solomon Katzenelbogen, "A Critical Appraisal of the Shock Therapies in the Major Psychoses and Psychoneuroses, III—Convulsive Therapy," *Psychiatry* 3 (1940):409–420; Leon Reznikoff, "Evaluation of Metrazol Shock in Treatment of Schizophrenia," *Archives of Neurology and Psychiatry* 43 (1940): 318–325; Richard Whitehead, "Pharmacologic and Pathologic Effects of Repeated Convulsant Doses of Metrazol," *American Journal of the Medical Sciences* 199 (1940):352–359; Lewis, "The Present Status of Shock Therapy of Mental Disorders"; Humbert, "Critique and Indications of Treatments in Schizophrenia."

50. Meduna, "General Discussion of the Cardiazol Therapy," 49–50.

51. William Menninger, "The Results with Metrazol as an Adjunct Therapy in Schizophrenia and Depressions," *Bulletin of the Menninger Clinic* 2 (1938): 129–141; Rankine Good, "Some Observations on the Psychological Aspects of Cardiazol Therapy," *Journal of Mental Science* 86 (1940):491–501; Rankine Good, "Convulsion Therapy in War Psychoneurotics," *Journal of Mental Science* 87 (1941):409–415.

52. Reznikoff, "Evaluation of Metrazol Shock," 325. Also see Menninger, "Results with Metrazol"; Kwalwasser, "Report on 441 Cases Treated with Metrazol."

53. Menninger, "Results with Metrazol"; A. J. Bain, "The Influence of Cardiazol on Chronic Schizophrenia," *Journal of Mental Science* 86 (1940):510–512; Good, "Some Observations on the Psychological Aspects of Cardiazol Therapy" and "Convulsion Therapy in War Psychoneurotics"; Bennett, *Fifty Years in Neurology and Psychiatry,* 131; Katzenelbogen, "A Critical Appraisal of the Shock Therapies in the Major Psychoses and Psychoneuroses, III—Convulsive Therapy," 419.

54. L. C. Cook, "Has Fear Any Therapeutic Significance in Convulsion Therapy?" *Journal of Mental Science* 86 (1940):484.

55. Roy Grinker, "Psychological Observations in Affective Psychoses Treated with Combined Convulsive Shock and Psychotherapy," *Journal of Nervous and Mental Disease* 97 (1943):623.

56. Abram Bennett, "Convulsive Shock Therapy in Depressive Psychoses," *American Journal of the Medical Sciences* 196 (1938):420–428.

57. Walter Freeman, "Brain-Damaging Therapeutics," *Diseases of the Nervous System* 2 (1940):83.

58. A. Warren Stearns, "Report on Medical Progress," *New England Journal of Medicine* 220 (1939):709–710.

59. As cited by Fink, "Meduna and the Origins of Convulsive Therapy."

60. Lucio Bini, "Experimental Researches on Epileptic Attacks Induced by the Electric Current," *American Journal of Psychiatry,* supplement 94 (1938):172–174.

61. As quoted by Frank, *History of Shock Treatment,* 9.

62. Ibid., 1–11; David Impastato, "The Story of the First Electroshock Treatment," *American Journal of Psychiatry* 116 (1960):1113–1114; Norman Endler, *Electroconvulsive Therapy: The Myths and the Realities* (Hans Huber Publishers, 1988), 18.

63. As quoted by Frank, *History of Shock Treatment,* 58.

64. Henry Stack Sullivan, "Explanatory Conceptions," *Psychiatry* 3 (1940):73.

65. Lewis, "The Present Status of Shock Therapy of Mental Disorders," 239.

66. Victor Gonda, "Treatment of Mental Disorders with Electrically Induced Convulsions," *Diseases of the Nervous System* 2 (1941):84–92.

67. Lothar Kalinowsky, "Organic Psychotic Syndromes Occurring During Electric Convulsive Therapy," *Archives of Neurology and Psychiatry* 53 (1945):269–273.

68. Freeman, "Brain-Damaging Therapeutics," 83.

69. Jan-Otto Ottosson, "Psychological or Physiological Theories of ECT," *International Journal of Psychiatry* 5 (1968):170–174.

70. Abraham Myerson, "Borderline Cases Treated by Electric Shock," *American Journal of Psychiatry* 100 (1943):353–357.

71. Kalinowsky, *Shock Treatments and Other Somatic Procedures,* 179.

72. Abram Bennett, "An Evaluation of the Shock Therapies," *Diseases of the Nervous System* 6 (1945):20–23; Abram Bennett, "Evaluation of Progress in Established Physiochemical Treatments in Neuropsychiatry," *Diseases of the Nervous System* 10 (1949):195–205.

73. Wellington Reynolds, "Electric Shock Treatment," *Psychiatric Quarterly* 19 (1945):322–333.

74. Upton published the account of her treatment, which relied on written medical records she obtained from Nazareth Sanatorium, in *Madness Network News,* in July 1975. Frank republished it in his *A History of Shock Treatment,* 64–67.

75. Lauretta Bender, "One Hundred Cases of Childhood Schizophrenia Treated with Electric Shock," *Transactions of the American Neurological Association* 72 (1947):165–168; E. R. Clardy, "The Effect of Electric Shock Treatment on Children Having Schizophrenic Manifestations," *Psychiatric Quarterly* 28 (1954):616–623.

76. Max Fink, "Experimental Studies of the Electroshock Process," *Diseases of the Nervous System* 19 (1958):113–117; Max Fink, "Effect of Anticholinergic Agent, Diethazine, on EEG and Behavior," *Archives of Neurology and Psychiatry* 80 (1958):380–386; Max Fink, "Cholinergic Aspects of Convulsive Therapy," *Journal of Nervous and Mental Disease* 142 (1966):475–481.

77. Franklin Ebaugh, "Fatalities Following Electric Convulsive Therapy: Report of Two Cases with Autopsy," *Archives of Neurology and Psychiatry* 49 (1943):107–117; Bernard Alpers, "The Brain Changes in Electrically Induced Convulsions in the Human," *Journal of Neuropathology and Experimental Neurology* 1 (1942):173–180; Lewis, "The Present Status of Shock Therapy of Mental Disorders," 239–240; James Huddleson, "Complications in Electric Shock Therapy," *American Journal of Psychiatry* 102 (1946):594–598; Salzman, "An Evaluation of Shock Therapy"; Albert Rabin, "Patients Who Received More Than One Hundred Electric Shock Treatments," *Journal of Personality* 17 (1948):42–48; Irving Janis, "Memory Loss Following Convulsive Treatments," *Journal of Personality* 17 (1948):29–32; Irving Janis, "Psychologic Effects of Electric Convulsive Treatments," *Journal of Nervous and Mental Disease* 111 (1950):359–382. Also see Donald Templer, "Cognitive Functioning and Degree of Psychosis in Schizophrenics Given Many Electroconvulsive Treatments," *British Journal of Psychiatry* 123 (1973):441–443; John Friedberg, "Shock Treatment, Brain Damage, and Memory Loss: A Neurological Perspective," *American Journal of Psychiatry* 134 (1977):1010–1018; and Peter Breggin, *Brain-Disabling Treatments in Psychiatry* (Springer Publishing Company, 1997), 129–157.

78. Deutsch, *The Shame of the States*, 161.

79. Kalinowsky, "Organic Psychotic Syndromes Occurring During Electric Convulsive Therapy"; Thelma Alper, "An Electric Shock Patient Tells His Story," *Journal of Abnormal and Social Psychology* 43 (1948):201–210; Seymour Fisher, "The Conscious and Unconscious Attitudes of Psychotic Patients Toward Electric Shock Treatment," *Journal of Nervous and Mental Disease* 118 (1953):144–149; Libby Blek, "Somatic Therapy as Discussed by Psychotic Patients," *Journal of Abnormal and Social Psychology* 50 (1955):394–400; Abse, "Transference and Countertransference in Somatic Therapies."

80. Ellen Field, *The White Shirts* (Tasmania Press, 1964), 6–7.

81. Sylvia Plath, *The Bell Jar* (Harper and Row, 1971), 160–161.

82. Dorothy Washburn Dundas, essay in *Beyond Bedlam*, ed. Jeanine Grob (Third Side Press, 1995), 34.

83. Donna Allison, letter to the editor, *Los Angeles Free Press*, April 18, 1969, as cited by Frank, *History of Shock Treatment*, 83.

84. Braslow, *Mental Ills and Bodily Cures*, 116.

85. Abram Bennett, "Evaluation of Progress in Established Physiochemical Treatments in Neuropsychiatry," *Diseases of the Nervous System* 10 (1949):195–205.

86. David Impastato, "Prevention of Fatalities in Electroshock Therapy," *Diseases of the Nervous System* 18, sec. 2 (1957):34–75.

87. Franklin Ebaugh, "A Review of the Drastic Shock Therapies in the Treatment of the Psychoses," *Annals of Internal Medicine* 18 (1943):279–296.

88. Fink, "Experimental Studies of the Electroshock Process."

89. Robert Jay Lifton, *The Nazi Doctors* (Basic Books, 1986), 299.

90. Braslow, *Mental Ills and Bodily Cures,* 105–106.

91. Abse, "Transference and Countertransference in Somatic Therapies"; Ruffin, "Attitudes of Auxiliary Personnel Administering Electroconvulsive and Insulin Coma Treatment."

92. As cited by Frank, *History of Shock Treatment,* 106.

93. Lewis Sharp, "Management of the Acutely Disturbed Patient by Sedative Electroshock Therapy," *Diseases of the Nervous System* 14 (1953):21–23.

94. Peter Cranford, *But for the Grace of God: The Inside Story of the World's Largest Insane Asylum. Millidgeville!* (Great Pyramid Press, 1981), 158.

Chapter 5: Brain Damage as Miracle Therapy

1. Freeman, "Brain-Damaging Therapeutics," 83.

2. Harold Himwich, "Electroshock: A Round Table Discussion," *American Journal of Psychiatry* 100 (1943):361–364.

3. As quoted by Elliott Valenstein, *Great and Desperate Cures* (Basic Books, 1986), 89. Also see Jack Pressman, *Last Resort* (Cambridge University Press, 1998), 50–53.

4. As quoted by Valenstein, *Great and Desperate Cures,* 89.

5. Carlyle Jacobsen, "Functions of Frontal Association Area in Primates," *Archives of Neurology and Psychiatry* 33 (1935):558–569; "Experimental Analysis of the Functions of the Frontal Association Areas in Primates," *Archives of Neurology and Psychiatry* 34 (1935):884–888; and "An Experimental Analysis of the Functions of the Frontal Association Areas in Primates," *Journal of Nervous and Mental Disease* 82 (1935):1–14.

6. As quoted by Walter Freeman and James Watts, *Psychosurgery* (Charles C. Thomas, 1950), xv.

7. Biographical material on Moniz from the following: Francisco Ruben Perino, "Egas Moniz, Founder of Psychosurgery, Creator of Angiography," *Journal of the International College of Surgeons* 36 (1961):261–271; Robert Wilkins, "Neurosurgical Classic," *Journal of Neurosurgery* (1964):1108–1109; Almeida Lima, "Egas Moniz 1987–1955," *Surgical Neurology* 1 (1973):247–248; Antonio Damasio, "Egas Moniz, Pioneer of Angiography and Leucotomy," *Mt. Sinai Journal of Medicine* 42 (1975):502–513; Valenstein, *Great and Desperate Cures,* 62–79; and Pressman, *Last Resort,* 53–54. Quotation is from Perino, "Egas Moniz, Founder of Psychosurgery, Creator of Angiography," 261.

8. As quoted by Valenstein, *Great and Desperate Cures,* 94.

9. Jacobsen, "An Experimental Analysis of the Functions of the Frontal Association Areas in Primates," 10.

10. Egas Moniz, "Essai d'un traitement chirurgical de certaines psychoses," *Bulletin de l'Academie de Medicine* 115 (1936):385–392. An English translation appears in *Journal of Neurosurgery* 21 (1964):1110–1114.

11. As quoted by Freeman and Watts, *Psychosurgery,* xvi.

12. See Valenstein, *Great and Desperate Cures*, 104.

13. Moniz, "Essai d'un traitement," trans. in *Journal of Neurosurgery*, 1113.

14. Walter Freeman, "Review of 'Tentative opératoires dans le traitement de certaines psychoses,'" *Archives of Neurology and Psychiatry* 36 (1936):1413.

15. See Valenstein, *Great and Desperate Cures*, 122–140; and Pressman, *Last Resort*, 71–77.

16. As quoted by Pressman, *Last Resort*, 75.

17. Freeman and Watts, *Pyschosurgery*, xviii–xix.

18. Walter Freeman, "Prefrontal Lobotomy in the Treatment of Mental Disorders," *Southern Medical Journal* 30 (1937):23–31; Walter Freeman, "Psychosurgery: Effect on Certain Mental Symptoms of Surgical Interruption of Pathways in the Frontal Lobe," *Journal of Nervous and Mental Disease* 88 (1938):587–601; Walter Freeman, "Some Observations on Obsessive Tendencies Following Interruption of the Frontal Association Pathways," *Journal of Nervous and Mental Disease* 88 (1938):224–234.

19. "Find New Surgery Aids Mental Cases," *New York Times*, November 21, 1936; "Surgery Used on the Soul-Sick; Relief of Obsessions Is Reported," *New York Times*, June 7, 1937. Also see Valenstein, *Great and Desperate Cures*, 154–155.

20. Freeman and Watts, *Psychosurgery*, 392–397; Freeman, "Psychosurgery: Effect on Certain Mental Symptoms," 595.

21. As quoted by Pressman, *Last Resort*, 78.

22. James Lyerly, "Prefrontal Lobotomy in Involutional Melancholia," *Journal of the Florida Medical Association* 25 (1938):225–229.

23. Ibid. The comments from Davis and Dodge are in the discussion section of this article.

24. The story of these two patients, Julia Koppendorf and Sally Gold, is reported by Pressman, *Last Resort*, 106–108.

25. As quoted in ibid., 142.

26. Edward Strecker, "A Study of Frontal Lobotomy," *American Journal of Psychiatry* 98 (1942):524–532.

27. Lloyd Ziegler, "Bilateral Prefrontal Lobotomy," *American Journal of Psychiatry* 100.1 (1943):178–184.

28–39. Freeman and Watts, *Psychosurgery*, 137–139, 29, 406, 157, 184, 185, 190, 198, 199, 195, 226–257, 257, 565–566.

40. Walter Freeman, "History of Development of Psychosurgery," *Digest of Neurology and Psychiatry* 17 (1949):412–451; quote is by David Rioch, 428.

41. See Pressman, *Last Resort*, 47–73, 86–101.

42. Editorial, "The Surgical Treatment of Certain Psychoses," *New England Journal of Medicine* 214 (1936):1088.

43. As cited by Pressman, *Last Resort*, 50.

44. As cited in ibid., 91.

45. Stanley Cobb, "Presidential Address," *Transactions of the American Neurological Association* (1949):1–7.

46. Panel discussion, "Neurosurgical Treatment of Certain Abnormal Mental States," *Journal of American Medical Association* 117 (1941):517–527.

47. As cited by Pressman, *Last Resort*, 108.

48. David Cleveland, "Prefrontal Lobotomy: Fifteen Patients Before and After Operation," *American Journal of Psychiatry* 101 (1945):749–755.

49. Freeman, "History of Development of Psychosurgery," 430–431.

50. Panel discussion, "Neurosurgical Treatment of Certain Abnormal Mental States," 518.

51. See Valenstein, *Great and Desperate Cures,* 157; *New York Times,* "Surgery Restores Incurably Insane," March 19, 1948.

52. Walter Kaempffert, "Turning the Mind Inside Out," *Saturday Evening Post,* May 24, 1941.

53. Freeman and Watts, *Psychosurgery,* 51–57. Also see Valenstein, *Great and Desperate Cures,* 199–220.

54. Freeman and Watts, *Psychosurgery,* 113.

55. Cranford, *But for the Grace of God,* 157.

56. Matthew Moore, "Some Experiences with Transorbital Leucotomy," *American Journal of Psychiatry* 107 (1951):801–807.

57. Braslow, *Mental Ills and Bodily Cures,* 125–151.

58. Ibid., 131, 139.

59. Ibid., 168–169.

60. Freeman and Watts, *Psychosurgery,* 436.

61. W. J. Mixter, "Frontal Lobotomy in Two Patients with Agitated Depression," *Archives of Neurology and Psychiatry* 44 (1940):236–239; quote is from discussion section.

62. "First International Congress on Psychosurgery," *American Journal of Psychiatry* 105 (1949):550–551.

63. Editorial, "Nobel Prize in Medicine," *New England Journal of Medicine* 241 (1949):1025.

64. "Explorers of the Brain," *New York Times* editorial, October 30, 1949.

Chapter 6: Modern-Day Alchemy

1. N. William Winkelman, Jr., "Chlorpromazine in the Treatment of Neuropsychiatric Disorders," *Journal of the American Medical Association* 155 (1954):18–21.

2. Shorter, *History of Psychiatry,* 255.

3. Pressman, *Last Resort,* 148; David Rothschild, *Diseases of the Nervous System* 11 (1951):147–150; D. Ewen Cameron, "Production of Differential Amnesia as a Factor in the Treatment of Schizophrenia," *Comprehensive Psychiatry* 1 (1960):26–33; Cameron, "The Depatterning Treatment of Schizophrenia," *Comprehensive Psychiatry* 3 (1962):65–76; Robitscher, *Eugenic Sterilization,* 123.

4. As cited by Judith Swazey, *Chlorpromazine in Psychiatry* (Massachusetts Institute of Technology Press, 1974), 105. For this early history, see Peter Breggin, *Toxic Psychiatry* (St. Martin's Press, 1991), and David Cohen, "A Critique of the Use of Neuroleptic Drugs," essay in *From Placebo to Panacea,* ed. Seymour Fisher and Roger Greenberg (John Wiley and Sons, 1997), 173–228.

5. Swazey, *Chlorpromazine in Psychiatry,* 134–135.

6. D. Anton-Stephens, "Preliminary Observations on the Psychiatric Uses of Chlorpromazine," *Journal of Mental Science* 199 (1954):543–557.

7. H. E. Lehmann, "Chlorpromazine: New Inhibiting Agent for Psychomotor Excitement and Manic States," *Archives of Neurology and Psychiatry* 71 (1954):227–237; Lehmann, "Therapeutic Results with Chlorpromazine in Psychiatric Conditions," *Canadian Medical Association Journal* 72 (1955):91–98; Lehmann, "Neurophysiologic Activity of Chlorpromazine in Clinical Use," *Journal of Clinical and Experimental Psychopathology* 17 (1956):129–141.

8. Winkelman, "Chlorpromazine in the Treatment of Neuropsychiatric Disorders."

9. Irvin Cohen, "Undesirable Effects and Clinical Toxicity of Chlorpromazine," *Journal of Clinical and Experimental Psychopathology* 17 (1956):153–162.

10. As cited by David Cohen, *From Placebo to Panacea*, 181.

11. "Chlorpromazine and Mental Health," Proceedings of the Symposium Held Under the Auspices of Smith, Kline & French Laboratories, June 6, 1955 (Lea and Febiger, 1955):51, 55, 73.

12. Ibid., 183.

13. Joel Elkes, "Effects of Chlorpromazine on the Behaviour of Chronically Overactive Psychotic Patients," *British Medical Journal* 2 (1954):560–565.

14. Lehmann, "Neurophysiologic Activity of Chlorpromazine in Clinical Use."

15. See ibid., 136.

16. "Chlorpromazine and Mental Health," 133, 158.

17. As cited by Breggin, *Toxic Psychiatry*, 73.

18. "Chlorpromazine and Mental Health," 86.

19. As cited by Ann Braden Johnson, *Out of Bedlam* (Basic Books, 1990), 26.

20. This history is detailed in Morton Mintz's *The Therapeutic Nightmare* (Houghton Mifflin, 1965), 70–92. The "sissy" quote is from the book's appendix, 488.

21. *Study of Administered Prices in the Drug Industry*, a report by the Senate Subcommittee on Antitrust and Monopoly, reprinted as an appendix in ibid., 477.

22. Ibid., 481.

23. As quoted by Shorter, *History of Psychiatry*, 253.

24. As quoted by Swazey, *Chlorpromazine in Psychiatry*, 190. See 159–207 for details on Smith Kline & French's marketing of Thorazine.

25. "Wonder Drug of 1954?" *Time*, June 14, 1954.

26. The dates of the 11 articles on chlorpromazine to appear in the *New York Times* in 1955 were: January 9; March 10; March 11; April 1; May 10; May 11; May 18; June 26; August 4; September 29; October 7. Also see in early 1956 articles on January 8, February 20, and February 25.

27. "Aid for the Mentally Ill," *New York Times*, June 26, 1955.

28. "Wonder Drugs: New Cure for Mental Ills?" *U.S. News and World Report*, June 17, 1955.

29. "Pills for the Mind," *Time*, March 7, 1955.

30. "2 Southerners Back Stevenson," *New York Times*, August 13, 1955.

31. Winkelman, "Chlorpromazine in the Treatment of Neuropsychiatric Disorders."

32. N. William Winkelman, Jr., "An Appraisal of Chlorpromazine," *American Journal of Psychiatry* 113 (1957):961–971.

33. "Drugs for the Mind," *Nation,* July 21, 1956.

34. "Analyst Hits Use of Calming Drugs," *New York Times,* March 11, 1956.

35. "Tranquilizer Drugs Are Held Harmful," *New York Times,* December 19, 1956.

36. "House Opens Inquiry on Tranquilizer Ads," *New York Times,* February 12, 1958; and "Tranquilizer Study Told of Curb on Ads," *New York Times,* February 13, 1958.

37. See Mintz, *The Therapeutic Nightmare,* 348–359.

38. Swazey, *Chlorpromazine in Psychiatry,* 161; also see Johnson, *Out of Bedlam,* 48.

39. Joint Commission on Mental Illness and Mental Health, *Action for Mental Health* (Science Editions, 1961):39.

40. "President Seeks Funds to Reduce Mental Illness," *New York Times,* February 6, 1963.

41. Henry Brill, "Analysis of 1955–1956 Population Fall in New York State Mental Hospitals in First Year of Large-Scale Use of Tranquilizing Drugs," *American Journal of Psychiatry* 114 (1957):509–517; "Analysis of Population Reduction in New York State Mental Hospitals During the First Four Years of Large Scale Therapy with Psychotropic Drugs," *American Journal of Psychiatry* 116 (1959) 495–508; and "Clinical-Statistical Analysis of Population Changes in New York State Mental Hospitals Since Introduction of Psychotropic Drugs," *American Journal of Psychiatry* 119 (1962):20–35.

42. Leon Epstein, "An Approach to the Effect of Ataraxic Drugs on Hospital Release Rates," *American Journal of Psychiatry* 119 (1962):36–47.

43. The National Institute of Mental Health Psychopharmacology Service Center Collaborative Study Group, "Phenothiazine Treatment in Acute Schizophrenia," *Archives of General Psychiatry* 10 (1964):246–261.

44. As cited by Grob, *The Mad Among Us,* 280.

Chapter 7: The Patients' Reality

1. Judi Chamberlin, *On Our Own* (McGraw-Hill, 1978), 52.

2. L. Farde, "Positron Emission Tomography Analysis of Central D_1 and D_2 Dopamine Receptor Occupancy in Patients Treated with Classical Neuroleptics and Clozapine," *Archives of General Psychiatry* 49 (1992):538–544. Also, G. P. Reynolds, "Antipsychotic Drug Mechanisms and Neurotransmitter Systems in Schizophrenia," *Acta Psychiatrica Scandinavica* 89, supplement 380 (1994):36–40.

3. Breggin, *Toxic Psychiatry,* 56.

4. Eric Kandel, ed., *Principles of Neural Science,* 3rd ed. (Elsevier, 1991), 863.

5. Pierre Deniker, "From Chlorpromazine to Tardive Dyskinesia: Brief History of the Neuroleptics," *Psychiatric Journal of the University of Ottawa* 14 (1989):253–259.

6. Mary Boyle, "Is Schizophrenia What It Was? A Re-Analysis of Kraepelin's and Blueler's Population," *Journal of the History of the Behavioral Sciences* 26 (1990):323–333. See also Mary Boyle, *Schizophrenia: A Scientific Delusion?* (Routledge, 1990).

7. Oliver Sacks, *Awakenings* (E. P. Dutton, 1973; paperback edition 1983), 13–23.

8. William Carpenter, "Treatment of Negative Symptoms," *Schizophrenia Bulletin* 11 (1985):440–449.

9. John Mirowsky, "Subjective Boundaries and Combinations in Psychiatric Diagnoses," *Journal of Mind and Behavior* 11 (1990):407–424.

10. R. E. Kendall, "Diagnostic Criteria of American and British Psychiatrists," *Archives of General Psychiatry* 25 (1971):123–130.

11. As cited by Seth Farber, *Madness, Heresy, and the Rumor of Angels* (Open Court, 1993), 190–240.

12. Alan Lipton, "Psychiatric Diagnosis in a State Hospital: Manhattan State Revisited," *Hospital and Community Psychiatry* 36 (1985):368–373; Heinz Lehmann, "Discussion: A Renaissance of Psychiatric Diagnosis?" *American Journal of Psychiatry* 125, supplement 10 (1969):43–46.

13. D. L. Rosenhan, "On Being Sane in Insane Places," *Science* 179 (1973): 250–258.

14. As cited by Herb Kutchins and Stuart Kirk, *Making Us Crazy* (Free Press, 1997), 205.

15. Samuel Cartwright, "Report on the Diseases and Physical Peculiarities of the Negro Race," *New Orleans Medical and Surgical Journal* 7 (1851):691–715.

16. J. M. Buchanan, "Insanity in the Colored Race," *New York Medical Journal* 44 (1886):67–70.

17. W. M. Bevis, "The Psychological Traits of the Southern Negro with Observations as to Some of His Psychoses," *American Journal of Psychiatry* 1 (1921):69–78. Franklin's story is told in Kirk, *Making Us Crazy*, 230–231.

18. Mary O'Malley, "Psychoses in the Colored Race," *American Journal of Insanity* 71 (1914):309–337.

19. Marti Loring, "Gender, Race, and DSM-III: A Study of the Objectivity of Psychiatric Diagnostic Behavior," *Journal of Health and Social Behavior* 29 (1988):1–22.

20. Edward Jarvis, *Insanity and Idiocy in Massachusetts: Report of the Commission on Lunacy, 1855* (Harvard University Press, 1971), 53.

21. C. E. Holzer, "The Increased Risk for Specific Psychiatric Disorders Among Persons of Low Socioeconomic Status," *American Journal of Social Psychiatry* 6 (1986):259–271.

22. Lipton, "Psychiatric Diagnosis in a State Hospital," 371.

23. Theodore van Putten: "Why Do Schizophrenic Patients Refuse to Take Their Drugs?" *Archives of General Psychiatry* 31 (1974):67–72; "Drug Refusal in Schizophrenia and the Wish to Be Crazy," *Archives of General Psychiatry* 33 (1976):1443–1446; and "Response to Antipsychotic Medication: The Doctor's and the Consumer's View," *American Journal of Psychiatry* 141 (1984):16–19. Also E. B. Larsen, "Subjective Experience of Treatment, Side Effects, Mental State and Quality of Life in Chronic Schizophrenics," *Acta Psychiatrica Scandinavica* 93 (1996):381–388.

24. Janet Gotkin, *Too Much Anger, Too Many Tears* (Quadrangle/The New York Times Book Co., 1975), 385; the longer quote is from Gotkin's testimony before the Senate. U.S. Senate, Committee on the Judiciary, Subcommittee to Investi-

gate Juvenile Delinquency, *Drugs in Institutions*, 94th Cong., 1st sess., 1975 (Readex depository 77-9118).

25. Hudson testified in person before Bayh's committee. The quotations from Daniel Eisenberg, Anil Fahini, and Beth Guiros were collected by the Network Against Psychiatric Assault and made part of the hearing record.

26. John Modrow, *How to Become a Schizophrenic* (Apollyon Press, 1992), 194–195.

27. Interview with Nathaniel Lehrman, October 1, 2000.

28. Robert Belmaker and David Wald, "Haloperidol in Normals," *British Journal of Psychiatry* 131 (1977):222–223.

29. Marjorie Wallace, "Schizophrenia—a National Emergency: Preliminary Observations on SANELINE," *Acta Psychiatrica Scandinavica* 89, supplement 380 (1994):33–35.

30. For Bayh's remarks, see vol. 3, p. 2, of the Senate subcommittee records, U.S. Senate, Committee on the Judiciary, Subcommittee to Investigate Juvenile Delinquency, *Drugs in Institutions*, 94th Cong., 1st sess., 1975.

31. Nina Schooler, "One Year After Discharge: Community Adjustment of Schizophrenic Patients," *American Journal of Psychiatry* 123 (1967):986–995.

32. Robert Prien, "Discontinuation of Chemotherapy for Chronic Schizophrenics," *Hospital and Community Psychiatry* 22 (1971):20–23.

33. George Gardos, "Maintenance Antipsychotic Therapy: Is the Cure Worse Than the Disease?" *American Journal of Psychiatry* 133 (1976):32–36; William Carpenter, Jr., "The Treatment of Acute Schizophrenia Without Drugs," *American Journal of Psychiatry* 134 (1977):14–20; Gerard Hogarty, "Fluphenazine and Social Therapy in the Aftercare of Schizophrenic Patients," *Archives of General Psychiatry* 36 (1979):1283–1294; George Gardos, "Withdrawal Syndromes Associated with Antipsychotic Drugs," *American Journal of Psychiatry* 135 (1978):1321–1324.

34. Interview with Sol Morris, February 2001 (Sol Morris is a pseudonym used by the patient).

35. J. Sanbourne Bockoven, "Comparison of Two Five-Year Follow-Up Studies: 1947 to 1952 and 1967 to 1972," *American Journal of Psychiatry* 132 (1975):796–801; Carpenter, "The Treatment of Acute Schizophrenia Without Drugs"; Maurice Rappaport, "Are There Schizophrenics for Whom Drugs May Be Unnecessary or Contraindicated?" *International Pharmacopsychiatry* 13 (1978):100–111; Susan Matthews, "A Non-Neuroleptic Treatment for Schizophrenia," *Schizophrenia Bulletin* 5 (1979):322–332.

36. Carpenter, "The Treatment of Acute Schizophrenia Without Drugs."

37. Guy Chouinard, "Neuroleptic-Induced Supersensitivity Psychosis," *American Journal of Psychiatry* 135 (1978):1409–1410; Chouinard, "Neuroleptic-Induced Supersensitivity Psychosis: Clinical and Pharmacologic Characteristics," *American Journal of Psychiatry* 137 (1980):16–20. Jones's quotation at the 1979 meeting of the Canadian Psychiatric Association is cited by Breggin in *Brain-Disabling Treatments in Psychiatry*, 60.

38. J. Sanbourne Bockoven, *Moral Treatment in American Psychiatry* (Springer Publishing, 1972); Nathaniel Lehrman, "A State Hospital Population Five Years After Admission," *Psychiatric Quarterly* 34 (1960):658–681; H. L. Rachlin, "Follow-

Up Study of 317 Patients Discharged from Hillside Hospital in 1950," *Journal of Hillside Hospital* 5 (1956):17–40; J. Sanbourne Bockoven, "Comparison of Two Five-Year Follow-Up Studies: 1947 to 1952 and 1967 to 1972"; Carpenter, "The Treatment of Acute Schizophrenia Without Drugs," and Rappaport, "Are There Schizophrenics for Whom Drugs May Be Unnecessary or Contraindicated?"

39. Peter Weiden, "The Cost of Relapse in Schizophrenia," *Schizophrenia Bulletin* 21 (1995):419–428; American Psychiatric Association, *Diagnostic and Statistical Manual of Mental Disorders*, 3rd ed. (APA, 1980).

40. Judith Godwin Rabkin, "Criminal Behavior of Discharged Mental Patients," *Psychological Bulletin* 86 (1979):1–27.

41. Jack Henry Abbott, *In the Belly of the Beast* (Vintage Books, 1991), 35–36.

42. M. Katherine Shear, "Suicide Associated with Akathisia and Depot Fluphenazine Treatment," *Journal of Clinical Psychopharmacology* 3 (1983):235–236; Theodore van Putten, "Phenothiazine-Induced Decompensation," *Archives of General Psychiatry* 30 (1974):102–105; Theodore van Putten, "The Many Faces of Akathisia," *Comprehensive Psychiatry* 16 (1975):43–46; Robert Drake, "Suicide Attempts Associated with Akathisia," *American Journal of Psychiatry* 142 (1985): 499–501; Theodore van Putten, "Behavioral Toxicity of Antipsychotic Drugs," *Journal of Clinical Psychiatry* 48 (1987):13–19.

43. Jerome Schulte, "Homicide and Suicide Associated with Akathisia and Haloperidol," *American Journal of Forensic Psychiatry* 6 (1985):3–7; Ed Shaw, "A Case of Suicidal and Homicidal Ideation and Akathisia in a Double-Blind Neuroleptic Crossover Study," *Journal of Clinical Psychopharmacology* 6 (1986):196–197; John Herrera, "High-Potency Neuroleptics and Violence in Schizophrenia," *Journal of Nervous and Mental Disease* 176 (1988):558–561; van Putten, "Behavioral Toxicity of Antipsychotic Drugs"; and Igor Galynker, "Akathisia as Violence," *Journal of Clinical Psychiatry* 58 (1997):31–32.

44. Nina Schooler, "Prevention of Relapse in Schizophrenia," *Archives of General Psychiatry* 37 (1980):16–24.

45. Theodore van Putten, "The Board and Care Home: Does It Deserve a Bad Press?" *Hospital and Community Psychiatry,* 30 (1979):461–464.

46. As cited by George Crane, "Tardive Dyskinesia in Patients Treated with Major Neuroleptics: A Review of the Literature," *American Journal of Psychiatry* 124, supplement (1968):40–47.

47. George Crane, "Clinical Psychopharmacology in Its 20th Year," *Science* 181 (1973):124–128; American Psychiatric Association, *Tardive Dyskinesia: A Task Force Report* (1992).

48. J. S. Paulsen, "Neuropsychological Impairment in Tardive Dyskinesia," *Neuropsychology* 8 (1994):227–241; John Waddington, "Cognitive Dysfunction in Schizophrenia: Organic Vulnerability Factor or State Marker for Tardive Dyskinesia?" *Brain and Cognition* 23 (1993):56–70; Michael Myslobodsky, "Central Determinants of Attention and Mood Disorder in Tardive Dyskinesia ('Tardive Dysmentia')," *Brain and Cognition* 23 (1993):88–101; R. Yassa, "Functional Impairment in Tardive Dyskinesia," *Acta Psychiatrica Scandinavica* 80 (1989):64–67.

49. G. Tsai, "Markers of Glutamergic Neurotransmission and Oxidative Stress Associated with Tardive Dyskinesia," *American Journal of Psychiatry* 155 (1998):

1207–1213; C. Thomas Gualtieri, "The Problem of Tardive Dyskinesia," *Brain and Cognition* 23 (1993):102–109; D. V. Jeste, "Study of Neuropathologic Changes in the Striatrum Following 4, 8 and 12 Months of Treatment with Fluphenazine in Rats," *Psychopharmacology* 106 (1992):154–160.

50. M. H. Chakos, "Increase in Caudate Nuclei Volumes of First-Episode Schizophrenic Patients Taking Antipsychotic Drugs," *American Journal of Psychiatry* 151 (1994):1430–1436; Raquel Gur, "Subcortical MRI Volumes in Neuroleptic-Naïve and Treated Patients with Schizophrenia," *American Journal of Psychiatry* 155 (1998):1711–1717.

51. Raquel Gur, "A Follow-Up Magnetic Resonance Imaging Study of Schizophrenia," *Archives of General Psychiatry* 55 (1998):145–152; Al Madsen, "Neuroleptics in Progressive Structural Brain Abnormalities in Psychiatric Illness," *Lancet* 352 (1998):784.

52. John Waddington, "Mortality in Schizophrenia," *British Journal of Psychiatry* 173 (1998):325–329; Louis Appleby, "Sudden Unexplained Death in Psychiatric In-Patients," *British Journal of Psychiatry* 176 (2000):405–406; and Javier Ballesteros, "Tardive Dyskinesia Associated with Higher Mortality in Psychiatric Patients," *Journal of Clinical Psychopharmacology* 20 (2000):188–194.

Chapter 8: The Story We Told Ourselves

1. John Modrow, *How to Become a Schizophrenic* (Apollyon Press, 1992), ix.

2. Susan Kemker, essay in *Pseudoscience in Biological Psychiatry*, ed. Colin Ross and Alvin Pam (John Wiley and Sons, 1995), 246.

3. Malcolm Bowers, "Central Dopamine Turnover in Schizophrenic Syndromes," *Archives of General Psychiatry* 31 (1974):50–54. In particular, see chart on 53.

4. Robert Post, "Cerebrospinal Fluid Amine Metabolites in Acute Schizophrenia," *Archives of General Psychiatry* 32 (1975):1063–1068; Francis White, "Differential Effects of Classical and Atypical Antipsychotic Drugs on A9 and A10 Dopamine Neurons," *Science* 221 (1983):1054–1056.

5. John Haracz, "The Dopamine Hypothesis: An Overview of Studies with Schizophrenic Patients," *Schizophrenia Bulletin* 8 (1982):438–458; Farouk Karoum, "Preliminary Evidence of Reduced Combined Output of Dopamine and Its Metabolites in Chronic Schizophrenia," *Archives of General Psychiatry* 44 (1987):604–607.

6. Tyrone Lee, "Binding of ^{3}H-Neuroleptics and ^{3}H-Apomorphine in Schizophrenic Brains," *Nature* 374 (1978):897–900; David Burt, "Antischizophrenic Drugs: Chronic Treatment Elevates Dopamine Receptor Binding in Brain," *Science* 196 (1977):326–328; Angus Mackay, "Increased Brain Dopamine and Dopamine Receptors in Schizophrenia," *Archives of General Psychiatry* 39 (1982):991–997; J. Kornhuber, "^{3}H-Siperone Binding Sites in Post-Mortem Brains from Schizophrenic Patients: Relationship to Neuroleptic Drug Treatment, Abnormal Movements, and Positive Symptoms," *Journal of Neural Transmission* 75 (1989):1–10.

7. John Kane, "Towards More Effective Antipsychotic Treatment," *British Journal of Psychiatry* 165, supplement 25 (1994):22–31.

8. Advertisement, "America's Pharmaceutical Research Companies," *New York Times Magazine*, August 18, 1996.

9. E. Fuller Torrey, *Surviving Schizophrenia: A Family Manual* (Harper and Row, 1983), 111.

10. Paul Keck, Jr., "Time Course of Antipsychotic Effects of Neuroleptic Drugs," *American Journal of Psychiatry* 146 (1989):1289–1292.

11. Calvin Turns, letter to the editor, *American Journal of Psychiatry* 147 (1990):1576.

12. Patricia Gilbert, "Neuroleptic Withdrawal in Schizophrenic Patients," *Archives of General Psychiatry* 52 (1995):173–188.

13. Ross Baldessarini, "Neuroleptic Withdrawal in Schizophrenic Patients," *Archives of General Psychiatry* 52 (1995):189–191; Adele Viguera, "Clinical Risk Following Abrupt and Gradual Withdrawal of Maintenance Neuroleptic Treatment," *Archives of General Psychiatry* 54 (1997):49–55.

14. Gerard Hogarty, "The Limitations of Antipsychotic Medication on Schizophrenia Relapse and Adjustment and the Contributions of Psychosocial Treatment," *Journal of Psychiatric Research* 32 (1998):243–250.

15. Weiden, "Cost of Relapse in Schizophrenia."

16. Torrey, *Surviving Schizophrenia*, 119.

17. George Crane, "Persistent Dyskinesia," *British Journal of Psychiatry* 122 (1973):395–405.

18. Nathan Kline, "On the Rarity of 'Irreversible' Oral Dyskinesias Following Phenothiazines," *American Journal of Psychiatry* 124 supplement (1968):48–54; Jonathan Cole, "Discussion of Dr. Crane's paper," *Transactions of New York Academy of Sciences* 36 (1974):658–661; John Curran, letter to the editor, *American Journal of Psychiatry* 130 (1973):1044; George Crane, "Clinical Psychopharmacology in Its 20th Year," *Science* 181 (1973):124–128. In *Toxic Psychiatry*, Peter Breggin writes in depth on psychiatry's resistance to recognizing this side effect and the financial influences behind that resistance.

19. U.S. Senate, Committee on the Judiciary, Subcommittee to Investigate Juvenile Delinquency, *Drugs in Institutions*, 94th Cong., 1st sess., 1975. See vols. 1 and 3. Also Thomas Gualtieri, "Preventing Tardive Dyskinesia and Preventing Tardive Dyskinesia Litigation," *Psychopharmacology Bulletin* 20 (1984):346–348.

20. Fred Gottlieb, "Report of the Speaker," *American Journal of Psychiatry* 142 (1985):1246–1249; "Report of the Treasurer," *American Journal of Psychiatry* 132 (1975):1109–1111.

21. Crane, "Clinical Psychopharmacology in Its 20th Year."

22. Daniel Freedman, "Editorial Comment," *Archives of General Psychiatry* 28 (1973):466–467.

23. There were an estimated 3 million Americans on neuroleptics by the early 1970s. The best estimate is that 3 percent to 5 percent of patients develop TD each year, which would be 90,000 to 150,000 people annually.

24. As cited by Breggin, *Toxic Psychiatry*, 79.

25. George Crane, "The Prevention of Tardive Dyskinesia," *American Journal of Psychiatry* 134 (1977):756–758.

26. Nancy Kennedy, "Disclosure of Tardive Dyskinesia: Effect of Written Policy on Risk Disclosure," *Psychopharmacology Bulletin* 28 (1992):93–100.

27. Gualtieri, "Preventing Tardive Dyskinesia and Preventing Tardive Dyskinesia Litigation."

28. Mantosh Dewan, "The Clinical Impact of the Side Effects of Psychotropic Drugs," in *The Limits of Biological Treatments for Psychological Distress,* ed. Seymour Fisher and Roger Greenberg (Lawrence Erlbaum Associates, 1989), 189–234.

29. Estimates of incidence rates for NMS vary from 0.2 percent to 1.4 percent. At a rate of 0.8 percent, that would mean approximately 24,000 cases annually from the 1960s to 1980s (with 3 million Americans on the drugs), with total deaths of 5,280 (24,000 x 22 percent mortality rate) annually. Over a twenty-year period, that would lead to more than 100,000 deaths. If the mortality rate had been 4 percent during that period, the number of deaths would have been reduced to approximately 20,000.

30. Deniker, "From Chlorpromazine to Tardive Dyskinesia."

31. Swazey, *Chlorpromazine in Psychiatry,* 130, 149; Swazey, *Chlorpromazine and Mental Health,* 154; H. A. Bowes, "Office Psychopharmacology," *Psychosomatics* 4 (1963) 22–26; Nathan Kline, "Psychopharmaceuticals: Uses and Abuses," *Postgraduate Medicine* 27 (1960):621.

32. Gerard Reardon, "Changing Patterns of Neuroleptic Dosage Over a Decade," *American Journal of Psychiatry* 146 (1989):726–729; Steven Segal, "Neuroleptic Medication and Prescription Practices with Sheltered-Care Residents: A 12-Year Perspective," *American Journal of Public Health* 82 (1992):846–852.

33. Deniker, "From Chlorpromazine to Tardive Dyskinesia," 259.

34. Gerard Hogarty, "Dose of Fluphenazine, Familial Expressed Emotion, and Outcome in Schizophrenia," *Archives of General Psychiatry* 45 (1988):797–805.

35. Frederic Quitkin, "Very High Dosage vs. Standard Dosage Fluphenazine in Schizophrenia," *Archives of General Psychiatry* 32 (1975):1276–1281; Thomas Hogan, "Pharmacotherapy and Suicide Risk in Schizophrenia," *Canadian Journal of Psychiatry* 28 (1983):277–281; Theodore van Putten, "A Controlled Dose Comparison of Haloperidol in Newly Admitted Schizophrenic Patients," *Archives of General Psychiatry* 47 (1990):754–758; Theodore van Putten, "Vulnerability to Extrapyramidal Side Effects," *Clinical Neuropharmacology* 6, supplement 1 (1983):S27–S34; and Herrera, "High-Potency Neuroleptics and Violence in Schizophrenia."

36. Deniker, "From Chlorpromazine to Tardive Dyskinesia," 258.

Chapter 9: Shame of a Nation

1. As cited by Roy Porter, *A Social History of Madness* (Weidenfeld and Nicolson, 1987).

2. Werner Tuteur, "The Discharged Mental Hospital Chlorpromazine Patient," *Diseases of the Nervous System* 20 (1959):512–517.

3. Frank Ayd, Jr., "The Depot Fluphenazines: A Reappraisal After 10 Years Clinical Experience," *American Journal of Psychiatry* 132 (1975):491–500. On noncompliance rates, see Peter Weiden, "Cost of Relapse in Schizophrenia," for summary of studies.

4. Copy is from a Smith Kline & French advertisement that ran monthly in *Mental Hospitals* in 1962.

5. See *Archives of General Psychiatry* (May 1974):562–563 for an example of this type of advertisement.

6. Heinz Lehmann, "The Philosophy of Long-Acting Medication in Psychiatry," *Diseases of the Nervous System* (1970):31, supplement 7–9.

7. Interview with David Oaks, January 2001.

8. See *New York Times,* May 3, 1982.

9. Abram Bennett's comments appeared in the *San Diego Union,* July 11, 1975; Alexander Rogawski's in the *Los Angeles Herald-Examiner,* May 30, 1976; both as cited in Frank's *History of Shock Treatment,* 111–112.

10. Thomas Gutheil, "In Search of True Freedom: Drug Refusal, Involuntary Medication, and 'Rotting with Your Rights On,'" *American Journal of Psychiatry* 137 (1980):327–328.

11. Taylor Branch, *Parting the Waters: America in the King Years, 1954–63* (Simon and Schuster, 1988), 344, 524.

12. "Abuse of Psychiatry for Political Repression in the Soviet Union," vol. 2, Committee on the Judiciary, U.S. Senate, 94th Cong., 2, 31.

13. Harvey Fireside, *Soviet Psychoprisons* (George J. McLeod, 1979), 147–148.

14. Ibid., 148.

15. Ibid., 125.

16. Ludmilla Thorne, "Inside Russia's Psychiatric Jails," *New York Times Magazine,* June 12, 1977.

17. David Ferleger, as cited in U.S. Senate, Committee on the Judiciary, Subcommittee to Investigate Juvenile Delinquency, *Drugs in Institutions,* 94th Cong., 1st sess., 1975, vol. 2, 170. Also see Maurice Deg. Ford, "The Psychiatrist's Double Bind: The Right to Refuse Medication," *American Journal of Psychiatry* 137 (1980):332–339.

18. "Mental Patients' Right to Refuse Medication Is Contested in Jersey," *New York Times,* March 28, 1981.

19. As quoted by Ford, "The Psychiatrist's Double Bind."

20. Paul Appelbaum, "The Right to Refuse Treatment with Antipsychotic Medications," *American Journal of Psychiatry* 145 (1988):413–419.

21. Interview with John Bola, December 2000.

22. "From Early Promise to Violent Death," *New York Times,* August 8, 1999.

23. Interview with Loren Mosher, December 1, 2000. Descriptions of patients, along with quotes from them, are taken from L. Mosher, "Soteria: Through Madness to Deliverance," unpublished manuscript, and from patient interviews in Susan Slanhoff's documentary, *Soterialand.*

24. Loren Mosher, "Community Residential Treatment for Schizophrenia: Two Year Followup," *Hospital and Community Psychiatry* 29 (1978):715–723; Susan Matthews, "A Non-Neuroleptic Treatment for Schizophrenia: Analysis of the Two-Year Postdischarge Risk of Relapse," *Schizophrenia Bulletin* 5 (1979):322–331; Mosher, "The Treatment of Acute Psychosis Without Neuroleptics: Six-Week Psychopathology Outcome Data from the Soteria Project," *International Journal of So-*

cial Psychiatry 41 (1995):157–173; Mosher, "The Soteria Project: Twenty-Five Years of Swimming Upriver," *Complexity and Change* 9 (2000):68–73.

25. Sources for this political battle include reviews by NIMH's "Clinical Program Projects Research Review Committee" on: April 27, 1970; April 1–2, 1973; April 1974; April 21, 1975; June 27, 1977; December 1, 1977; February 17–18, 1978; and June 26–27, 1978; and review by NIMH's "Mental Health Services Development Committee" in January 1974.

26. Luc Ciompi, "The Pilot Project Soteria Berne," *British Journal of Psychiatry* 161, supplement 18 (1992):145–153.

27. "Prozac's Worst Enemy," *Time,* October 10, 1994.

28. J. Leff, "The International Pilot Study of Schizophrenia: Five-Year Follow-Up Findings," *Psychological Medicine* 22 (1992):131–145. The outcomes described here are from fig. 1, 136, and table 5, 137. The best outcomes group at five years is composed of people who had one or more psychotic episodes but each time had a complete remission and were asymptomatic at the end of five years. The "so-so" group is composed of people who had a psychotic episode but never attained a complete remission. The worst outcomes group is composed of people who had multiple or continual psychotic episodes and never attained a complete remission.

29. Assen Jablensky, "Schizophrenia: Manifestations, Incidence and Course in Different Cultures, a World Health Organization Ten-Country Study," *Psychological Medicine,* supplement 20 (1992):1–95.

30. Ibid., 1.

31. Ibid., 60.

32. Courtenay Harding, "The Vermont Longitudinal Study of Persons with Severe Mental Illness," *American Journal of Psychiatry* 144 (1987):727–734; Harding, "Chronicity in Schizophrenia: Fact, Partial Fact, or Artifact?" *Hospital and Community Psychiatry* 38 (1987):477–485; Harding, "Empirical Correction of Seven Myths About Schizophrenia with Implications for Treatment," *Acta Psychiatrica Scandinavica* 384, supplement (1994):140–146. Also see Patrick McGuire, "New Hope for People with Schizophrenia," *APA Monitor* 31 (February 2000).

33. Interview with David Cohen, February 2001.

34. Anthony Lehman, "Patterns of Usual Care for Schizophrenia: Initial Results from the Schizophrenia Patient Outcomes Research Team Client Survey," *Schizophrenia Bulletin* 24 (1998):11–20.

Chapter 10: The Nuremberg Code Doesn't Apply Here

1. Interview with Shalmah Prince, September 1998.

2. "State Hospital Accused of Wrong Diagnoses, Fueling Debate over Nation's Mental Care," *New York Times,* April 23, 1985.

3. As cited in George Annas and Michael Grodin, eds., *The Nazi Doctors and the Nuremberg Code* (Oxford University Press, 1992), ix.

4. Paul Hoch, "Experimentally Produced Psychoses," *American Journal of Psychiatry* 107 (1951):607–611; Paul Hoch, "Effects of Mescaline and Lysergic Acid," *American Journal of Psychiatry* 108 (1952):579–584; and Nolan Lewis and Margaret Strahl, eds., *The Complete Psychiatrist* (State University of New York Press, 1968), 375–382.

5. Hoch, "Experimentally Produced Psychoses."

6. Paul Hoch, "Theoretical Aspects of Frontal Lobotomy and Similar Brain Operations," *American Journal of Psychiatry* 106 (1949):448–453.

7. Hoch, "Experimentally Produced Psychoses."

8. Elliot Luby, "Model Psychoses and Schizophrenia," *American Journal of Psychiatry* 118 (1962):61–67.

9. Leo Hollister, "Drug-Induced Psychoses and Schizophrenic Reactions: A Critical Comparison," *Annals of New York Academy of Sciences* 96 (1962):80–88.

10. David Janowsky, "Provocation of Schizophrenic Symptoms by Intravenous Adminstration of Methylphenidate," *Archives of General Psychiatry* 28 (1973): 185–191; David Janowsky, "Proceedings: Effect of Intravenous D-Amphetamine, L-Amphetamine and Methylphenidate in Schizophrenics," *Psychopharmacology Bulletin* 10 (1974):15–24; David Janowsky, "Methylphenidate, Dextroamphetamine, and Levamfetamine: Effects on Schizophrenic Symptoms," *Archives of General Psychiatry* 33 (1976):304–308.

11. Jeffrey Lieberman, "Brain Morphology, Dopamine, and Eye-Tracking Abnormalities in First-Episode Schizophrenia," *Archives of General Psychiatry* 50 (1993):357–368; Lieberman, "Time Course and Biologic Correlates of Treatment Response in First-Episode Schizophrenia," *Archives of General Psychiatry* 50 (1993):369–376.

12. Stephen Strakowski, "Lack of Enhanced Response to Repeated D-Amphetamine Challenge in First-Episode Psychosis," *Biological Psychiatry* 42 (1997):749–755.

13. Rajiv Sharma, "Behavioral and Biochemical Effects of Methylphenidate in Schizophrenic and Nonschizophrenic Patients," *Biological Psychiatry* 30 (1991): 459–466.

14. Anand Pandurangi, "Amphetamine Challenge Test, Response to Treatment, and Lateral Ventricle Size in Schizophrenia," *Biological Psychiatry* 25 (1989):207–214.

15. Michael Davidson, "L-Dopa Challenge and Relapse in Schizophrenia," *American Journal of Psychiatry* 144 (1987):934–938.

16. Peter Lucas, "Dysphoria Associated with Methylphenidate Infusion in Borderline Personality Disorder," *American Journal of Psychiatry* 144 (1987):1577–1579.

17. John Krystal, "M-Chlorophenylpiperazine Effects in Neuroleptic-Free Schizophrenic Patients," *Archives of General Psychiatry* 50 (1993):624–635.

18. A. K. Malhotra, "Ketamine-Induced Exacerbation of Psychotic Symptoms and Cognitive Impairment in Neuroleptic-Free Schizophrenics," *Neuropsychopharmacology* 17 (1997):141–150.

19. A. C. Lahti, "Subanesthetic Doses of Ketamine Stimulate Psychosis in Schizophrenia," *Neuropsychopharmacology* 13 (1995):9–19.

20. Interview with Vera Sharav, October 1998.

21. Interview with David Shore, September 1998.

22. Interview with Paul Appelbaum, September 1998.

23. Interview with Stephen Strakowski, September 1998.

24. Consent form for L-dopa study, obtained by Vera Sharav on March 1, 1994, from the Bronx VA Medical Center.

25. Letter from Dr. Michael Davidson to Vera Hassner (Sharav) on May 16, 1994.

26. Adam Wolkin, "Acute d-Amphetamine Challenge in Schizophrenia," *Biological Psychiatry* 36 (1994):317–325; consent form for study obtained through FOI request.

27. Consent form for study titled "NMDA Receptor Agonist Effect in Schizophrenia," dated September 1993.

28. Transcript of National Bioethics Advisory Meeting, May 19, 1998, at Case Western Reserve University in Cleveland, Ohio.

29. Interview with Michael Susko, October 1998

30. Interview with Franklin Marquit, October 1998

31. Interview with Wesley Alcorn, October 1998.

32. Court documents reviewed for this account include the informed consent signed by Shalmah Hawkins (Prince); Prince's medical records during her three weeks in the hospital; depositions by Jack Hirschowitz and David Garver; and the court's decision dismissing her case. Also, interviews with Shalmah Prince, August 1998 and August 2000.

33. Interview with Ken Faller, October 1998.

Chapter 11: Not So Atypical

1. *Action for Mental Health: Final Report of the Joint Commission on Mental Illness and Health, 1961* (Basic Books, 1961), 189.

2. T. Lewander, "Neuroleptics and the Neuroleptic-Induced Deficit Syndrome," *Acta Psychiatrica Scandinavica* 89, supplement 380 (1994):8–13; Peter Weiden, "Atypical Antipsychotic Drugs and Long-Term Outcome in Schizophrenia," *Journal of Clinical Psychiatry* 57, supplement 11 (1996):53–60; Richard Keefe, "Do Novel Antipsychotics Improve Cognition? A Report of a Meta-Analysis," *Psychiatric Annals* 29 (1999):623–629; George Arana, "An Overview of Side Effects Caused by Typical Antipsychotics," *Journal of Clinical Psychiatry* 61, supplement 8 (2000):5–13; William Glazer, "Review of Incidence Studies of Tardive Dyskinesia Associated with Atypical Antipsychotics," *Journal of Clinical Psychiatry* 61, supplement 4 (2000):15–25.

3. Weiden, "Atypical Antipsychotic Drugs and Long-Term Outcome in Schizophrenia"; Bruce Kinon, "Treatment of Neuroleptic-Resistant Schizophrenic Relapse," *Psychopharmacology Bulletin* 29 (1993):309–314.

4. Caleb Adler, "Comparison of Ketamine-Induced Thought Disorder in Healthy Volunteers and Thought Disorder in Schizophrenia," *American Journal of Psychiatry* 156 (1999):1646–1648.

5. Weiden, "Atypical Antipsychotic Drugs and Long-Term Outcome in Schizophrenia"; Arana, "An Overview of Side Effects Caused by Typical Antipsychotics" (quotations are from discussion section); Philip Harvey, "Cognitive Impairment in Schizophrenia: Its Characteristics and Implications," *Psychiatric Annals* 29 (1999):657–660; the patient-satisfaction data was presented at the Twelfth Congress of the European College of Neuropsychopharmacology, September 21–25, 1999. Survey was conducted by Walid Fakhouri, director of research at SANE, a British mental health group.

6. Stephen Marder, "Risperidone in the Treatment of Schizophrenia," *American Journal of Psychiatry* 151 (1994):825–835; Guy Chouinard, "A Canadian Multicenter Placebo-Controlled Study of Fixed Doses of Risperidone and Haloperidol in the Treatment of Chronic Schizophrenic Patients," *Journal of Clinical Psychopharmacology* 13 (1993):25–40. For Janssen advertisement touting risperidone's EPS as being as safe as a placebo, see *American Journal of Psychiatry* 151 (April 1994).

7. "New Hope for Schizophrenia," *Washington Post*, February 16, 1993; "Seeking Safer Treatments for Schizophrenia," *New York Times*, January 15, 1992.

8. Harvey, "Cognitive Impairment in Schizophrenia," 659.

9. See Charles Beasley, "Efficacy of Olanzapine: An Overview of Pivotal Clinical Trials," *Journal of Clinical Psychiatry* 58, supplement 10 (1997):7–12.

10. "Psychosis Drug from Eli Lilly Racks Up Gains," *Wall Street Journal*, April 14, 1998; "A New Drug for Schizophrenia Wins Approval from the FDA," *New York Times*, October 2, 1996.

11. Sales figures are from IMS Health; "Schizophrenia, Close-Up of the Troubled Brain," *Parade*, November 21, 1999; "Mental Illness Aid," *Chicago Tribune*, June 4, 1999; "Lives Recovered," *Los Angeles Times*, January 30, 1996. Quote is from headline.

12. I reported on the growth of the clinical trials industry for four years, 1994–1998, for a publishing company I co-founded, CenterWatch. Information is from *CenterWatch* reports 1994–1997; Peter Vlasses's quote is from *CenterWatch*, "Major Medical Centers Scramble for Clinical Grants," June 1, 1994.

13. Interview with Andrew Gottesman, "Cracks in the Partnership," *CenterWatch*, October 1995; other information is from *CenterWatch* articles: "Practice-Based Sites Prosper," March 1998; "Dedicated Sites Buoyed by Hot Market," April 1998; "The Top Ten Stories of 1997," January 1998; "Phymatrix Acquires Clinical Studies in Blockbuster Deal," June 1997; "Collaborative Clinical Nets $42 Million in IPO," August 1996. See also *Neuropractice* 7 (6) (2000):41, 43.

14. Marcia Angell, "Is Academic Medicine for Sale?" *New England Journal of Medicine* 342 (May 18, 2000):1516–1518.

15. Alan Gelenberg, "The Editor Responds," *Journal of Clinical Psychiatry* 60 (1999):122.

16. Thomas Bodenheimer, "Uneasy Alliance: Clinical Investigators and the Pharmaceutical Industry," *New England Journal of Medicine* 342 (May 18, 2000):1539–1544.

17. Angell, "Is Academic Medicine for Sale?" Angell spoke at the Third National Ethics Conference, sponsored by Friends Research Institute, November 4, 2000, Baltimore, MD.

18. Richard Borison, "Risperidone: Clinical Safety and Efficacy in Schizophrenia," *Psychopharmacology Bulletin* 28 (1992):213–218; "Seeking Safer Treatments for Schizophrenia," *New York Times*, January 15, 1992.

19. See "Drug Makers Relied on Clinical Researchers Who Now Await Trial," *Wall Street Journal*, August 15, 1997, for information on Borison's faking of trial data in the Thorazine study.

20. Information on Borison and Diamond is from a written report by the Department of Veterans Affairs, July 17, 1996, which includes transcripts of depositions from VA employees, and from the Bill of Indictment, State of Georgia, February 18, 1997.

21. Richard Borison, "Recent Advances in the Pharmacotherapy of Schizophrenia," *Harvard Review of Psychiatry* 4 (1997):255–271.

22. Information on deaths and suicides for the risperidone, olanzapine, and quetiapine trials was obtained from FDA documents, specifically the FDA's reviews of New Drug Applications (NDAs) for each of those drugs. Information on deaths and suicides in the sertindole trials was obtained from a transcript of the Forty-Seventh Meeting of the Psychopharmacological Drugs Advisory Committee, July 15, 1996.

23. Interview with Robert Temple, September 1998.

24. Interview with Ed Endersbe, September 1998.

25. The record of Faruk Abuzzahab's treatment of Susan Endersbe is from a record of "facts" stipulated to by Abuzzahab as part of his agreement with the Minnesota Board of Medical Practice, December 13, 1997. Abuzzahab's license to practice was temporarily suspended as part of that order.

26. Interview with Ross Baldessarini, September 1998. Baldessarini makes this same point in "Neuroleptic Withdrawal in Schizophrenic Patients," *Archives of General Psychiatry* 52 (1995):189–191.

27. FDA reviews of risperidone data included the following written commentaries: reviews by Andrew Mosholder, May 11, 1993 and November 7, 1993; David Hoberman, April 20, 1993; and Thomas Laughren, December 20, 1993.

28. Joseph McEvoy, "Optimal Dose of Neuroleptic in Acute Schizophrenia," *Archives of General Psychiatry* 48 (1991):739–745; van Putten, "Behavioral Toxicity of Antipsychotic Drugs," and "Controlled Dose Comparison of Haloperidol in Newly Admitted Schizophrenic Patients."

29. Internal memorandum from Paul Leber to Robert Temple, December 21, 1993. Obtained via FOI request.

30. Approval letter from Robert Temple to Janssen Research Foundation, December 29, 1993. Obtained via FOI request.

31. Stephen Marder, "The Effects of Risperidone on the Five Dimensions of Schizophrenia Derived By Factor Analysis: Combined Results of the North American Trials," *Journal of Clinical Psychiatry* 58 (1997):538–546.

32. Patricia Rosebush, "Neurologic Side Effects in Neuroleptic-Naïve Patients Treated with Haloperidol or Risperidone," *Neurology* 52 (1999):782–785.

33. Michael Knable, "Extrapyramidal Side Effects with Risperidone and Haloperidol at Comparable D_2 Receptor Levels," *Psychiatry Research: Neuroimaging Section* 75 (1997):91–101.

34. John Sweeney, "Adverse Effects of Risperidone on Eye Movement Activity: A Comparison of Risperidone and Haloperidol in Antipsychotic-Naïve Schizophrenic Patients," *Neuropsychopharmacology* 16 (1997):217–228.

35. Cameron Carter, "Risperidone Use in a Teaching Hospital During Its First Year After Market Approval: Economic and Clinical Implications," *Psychopharmacology Bulletin* 31 (1995):719–725; Renee Binder, "A Naturalistic Study of Clinical

Use of Risperidone," *Psychiatric Services* 49 (1998):524–526; Jeffrey Mattes, "Risperidone: How Good Is the Evidence for Efficacy?" *Schizophrenia Bulletin* 23 (1997):155–161.

36. Patricia Huston, "Redundancy, Disaggregation, and the Integrity of Medical Research," *Lancet* 347 (1996):1024–1026; Richard Horton, "Prizes, Publications, and Promotion," *Lancet* 348 (1996):1398.

37. "A New Drug for Schizophrenia Wins Approval from the FDA," *New York Times,* October 2, 1996.

38. FDA reviews of olanzapine data included the following written commentaries: reviews by Thomas Laughren on September 27, 1996; by Paul Andreason on July 29 and September 26, 1996; and by Paul Leber on August 18 and August 30, 1996.

39. Robert Conley, "Adverse Events Related to Olanzapine," *Journal of Clinical Psychiatry* 61, supplement 8 (2000):26–30.

40. FDA reviews of quetiapine data included the following written commentaries: reviews by Andrew Mosholder on June 13 and August 19, 1997; Thomas Laughren on August 21, 1997; and Paul Leber on September 24, 1997.

41. In their published articles, researchers at times anticipated criticism that the trials were biased by design and sought to answer the criticism before it was aired. For instance, Janssen-funded researchers argued that a 20 mg. dose of haloperidol was appropriate, despite the fact that dosing studies had found that it caused a high incidence of akathisia and other adverse side effects. See Guy Chouinard, "A Canadian Multicenter Placebo-Controlled Study of Fixed Doses of Risperidone and Haloperidol in the Treatment of Chronic Schizophrenic Patients," *Journal of Clinical Psychopharmacology* 13 (1993):25–40.

42. John Geddes, "Atypical Antipsychotics in the Treatment of Schizophrenia: Systematic Overview and Meta-Regression Analysis," *British Medical Journal* 321 (2000):1371–1376.

43. Pierre Tran, "Double-Blind Comparison of Olanzapine Versus Risperidone in the Treatment of Schizophrenia and Other Psychotic Disorders," *Journal of Clinical Psychopharmacology* 17 (1997):407–418; Criticism by Janssen that the study was biased, and Eli Lilly's response to that criticism, appeared in letters to the editor, *Journal of Clinical Psychopharmacology* 18 (1998):174–179; Ric Procyshyn, "Drug Utilization Patterns and Outcomes Associated with In-Hospital Treatment with Risperidone or Olanzapine," *Clinical Therapeutics* 20 (1998):1203–1217; B. C. Ho, "A Comparative Effectiveness Study of Risperidone and Olanzapine in the Treatment of Schizophrenia," *Journal of Clinical Psychiatry* 60 (1999):658–663.

44. "A Choice Between Treatment and Tragedy," *The Washington Post,* July 29, 1998.

45. Peter Weiden, *Breakthroughs in Antipsychotic Medications* (W. W. Norton, 1999), 26, 29, 63.

46. S. Silvestri, "Increased Dopamine D_2 Receptor Binding After Long-Term Treatment with Antipsychotics in Humans: A Clinical PET Study," *Psychopharmacology* 152 (2000):174–180.

47. Nancy Andreasen, "Understanding Schizophrenia: A Silent Spring?" *American Journal of Psychiatry* 155 (1998):1657–1659.

48. "Advocates Alarmed by Drugs Used for Kids," *Miami Herald,* May 7, 2001; "Radical Study on Schizophrenia May Be Expanded," *Wall Street Journal,* July 26, 2000.

Epilogue

1. Andrew Scull, "Cycles of Despair," *Journal of Mind and Behavior* 11 (1990):301–312.

2. Ben Thornley, "Content and Quality of 2000 Controlled Trials in Schizophrenia over 50 Years," *British Medical Journal* 317 (1998):1181–1184.

3. "Makers of Drugs for Mentally Ill Find East Asia is Resistant Market," *Wall Street Journal,* January 10, 2001.

INDEX

3/02